The Mathematical
Education of Teachers II

CBMS
Conference Board of the Mathematical Sciences

Issues in Mathematics Education

Volume 17

The Mathematical Education of Teachers II

American Mathematical Society
Providence, Rhode Island
in cooperation with
Mathematical Association of America
Washington, D. C.

The work of preparing the MET II report was made possible by a grant from Math for America.

2010 *Mathematics Subject Classification.* Primary 97A99, 00-01.

Library of Congress Cataloging-in-Publication Data

The mathematical education of teachers II.
    pages cm. – (Issues in mathematics education / CBMS, Conference Board of the Mathematical Sciences ; volume 17)
Includes bibliographical references.
ISBN 978-0-8218-6926-0 (alk. paper)
    1. Mathematics–Study and teaching–United States. 2. Mathematics teachers–Training of–United States. I. Conference Board of the Mathematical Sciences. II. Title: Mathematical education of teachers 2.

QA13.M3535   2012
510.71′1–dc23

                                                                        2012034726

About the Conference Board of the Mathematical Sciences

The Conference Board of the Mathematical Sciences (CBMS) is an umbrella organization consisting of sixteen professional societies all of which have as one of their primary objectives the increase or diffusion of knowledge in one or more of the mathematical sciences. Its purpose is to promote understanding and cooperation among these national organizations so that they work together and support each other in their efforts to promote research, improve education, and expand the uses of mathematics.

The CBMS member societies are:

American Mathematical Association of Two-Year Colleges

American Mathematical Society

Association of Mathematics Teacher Educators

American Statistical Association

Association for Symbolic Logic

Association for Women in Mathematics

Association of State Supervisors of Mathematics

Benjamin Banneker Association

Institute of Mathematical Statistics

Mathematical Association of America

National Association of Mathematicians

National Council of Supervisors of Mathematics

National Council of Teachers of Mathematics

Society for Industrial and Applied Mathematics

Society of Actuaries

TODOS: Mathematics for ALL

For more information about CBMS and its member societies, see `www.cbmsweb.org`.

Writing Team

Sybilla Beckmann, University of Georgia (lead writer, elementary grades chapter)

Daniel Chazan, University of Maryland

Al Cuoco, Education Development Center

Francis (Skip) Fennell, McDaniel College

Bradford Findell, The Ohio State University

Cathy Kessel, Mathematics Education Consultant (editor)

Karen King, National Council of Teachers of Mathematics

W. James Lewis, University of Nebraska–Lincoln (chair)

William McCallum, University of Arizona (lead writer, high school chapter)

Ira Papick, University of Nebraska–Lincoln

Barbara Reys, University of Missouri

Ronald Rosier, Conference Board of the Mathematical Sciences

Richard Scheaffer, University of Florida

Denise A. Spangler, University of Georgia (lead writer, middle grades chapter)

Alan Tucker, State University of New York at Stony Brook (lead writer, Chapters 1–3)

Contents

Preface

This report is a resource for those who teach mathematics—and statistics[1]—to PreK–12 mathematics teachers, both future teachers and those who already teach in our nation's schools. The report makes recommendations for the mathematics that teachers should know and how they should come to know that mathematics. It urges greater involvement of mathematicians and statisticians in teacher education so that the nation's mathematics teachers have the knowledge, skills, and dispositions needed to provide students with a mathematics education that ensures high school graduates are college- and career-ready as envisioned by the Common Core State Standards.

Mathematics teacher education is a complex, interdisciplinary enterprise requiring knowledge of teaching and learning as well as knowledge of mathematics. This argues strongly for a partnership between mathematics educators and those who teach mathematics. Thus, this report will also be an important resource for mathematics educators.

The *Mathematical Education of Teachers* (referred to as MET I in this report) was published in 2001. Since that time much has changed. In particular, the attention given by the mathematics profession to the mathematical education of teachers has increased as more mathematicians and statisticians have taken increasingly active roles in teacher preparation and content-based professional development for current teachers. The Math Science Partnerships (supported by the National Science Foundation and the United States Department of Education) and the NSF's Robert Noyce Teacher Scholarship Program have connected institutions of higher education with K–12 school systems, fostering new partnerships and extending existing collaborations. These and other changes in institutional support and emphasis have helped to increase the engagement of collegiate mathematicians and statisticians in teacher education.

Their engagement has proved to have a wide variety of benefits. For the mathematicians and statisticians, preparation and professional development for teachers can be genuinely interesting intellectual experiences, affording the opportunity to "think deeply about simple things," and to make connections between the undergraduate courses that they teach and K–12 mathematics.

Attending to the needs of prospective teachers by focusing on reasoning and proof across the spectrum of undergraduate mathematics courses that they take, helps them to make sense of mathematics—and makes it easier to understand, easier

---

[1]In K–12 schools, statistics is part of the mathematics curriculum. At the collegiate level, statistics is recognized as part of the mathematical sciences, but a separate discipline and most research universities have a separate department of statistics. Strengthening PreK–12 mathematics education requires the active involvement of both mathematicians and statisticians.

to teach, and intellectually satisfying for all course-takers. Thus, attending to the needs of future teachers in this way benefits all undergraduates.

For practicing K–12 teachers, content-based professional development offered by Math Science Partnerships has changed their attitudes about mathematics, and increased their mathematical interest and abilities. Moreover, it has increased the achievement of their students.

Determining the most important mathematics that teachers should know requires a clear vision of the mathematics that they will be expected to teach. The Common Core State Standards represent such a vision. Because most states have adopted the Common Core, the recommendations of this report focus on enabling teachers to teach that mathematics.

This report (MET II) draws on the experience and knowledge of the past decade to:

- Update MET I's recommendations for the mathematical preparation of teachers at all grade levels: elementary, middle, and high school.

- Address the professional development of teachers of mathematics.

- Discuss the mathematical needs of teachers with special responsibilities such as elementary mathematics specialists and special education teachers.

At the same time, MET II reiterates and elaborates themes of the first MET report:

- There is intellectual substance in school mathematics.

- Proficiency with school mathematics is necessary but not sufficient mathematical knowledge for a teacher.

- The mathematical knowledge needed for teaching differs from that of other professions.

- Mathematical knowledge for teaching can and should grow throughout a teacher's career.

Chapter 1 describes these themes in more detail, outlining the mathematical issues that underlie the recommendations in this report, including the structure and content of the Common Core. Chapter 2 summarizes empirical findings that underlie these recommendations and connects them with the current educational context. Chapter 3 gives recommendations for strengthening the mathematical education of teachers in the United States, with respect to the mathematics that teachers should learn and the roles of mathematicians and statisticians in their learning. This chapter will be of special interest to department chairs, policy-makers, and others in leadership positions.

Chapters 4, 5, and 6 give recommendations for the mathematical preparation and professional development of elementary, middle grades, and high school teachers. These will be the chapters of greatest importance for those engaged in teacher preparation or professional development.

Appendix A gives a short annotated list with two types of entries: recent reports whose conclusions inform the recommendations in this document, and sources of information about accreditation and licensure.

The Common Core State Standards have two categories: those concerning mathematical content and those concerning mathematical practice. Appendix B gives an overview of the content standards. The Standards for Mathematical Practice are given in Appendix C.

Web resources. Web resources associated with this report are located on the web site of the Conference Board of the Mathematical Sciences, `www.cbmsweb.org`. These are intended as an initial collection of relevant information rather than as a continuously updated reference.

Audience. This report should be useful to the entire community of professionals who educate teachers of mathematics, from those who teach undergraduates seeking initial certification to those who work with veteran teachers pursuing opportunities for professional development. Its audience includes professional development providers housed outside of academic institutions as well as collegiate faculty from disciplines outside the mathematical sciences who have become actively engaged in the mathematical education of teachers.

Its primary audiences, however, are faculty who teach in mathematics or statistics departments and their colleagues in colleges of education who have primary responsibility for the mathematical education of teachers. In addition, this report will be useful to policy-makers at all levels who look to the mathematics and mathematics education community for professional guidance with respect to the mathematical education of teachers. Thus, the three main audiences are:

> *Mathematicians and statisticians.* Faculty members of mathematics and statistics departments at two- and four-year collegiate institutions teach the mathematics and statistics courses taken by prospective and practicing teachers. Their departmental colleagues set policies regarding mathematics teacher preparation. At the risk of oversimplification, this report will at times refer to this audience as "mathematicians" or "mathematics faculty."

> *Mathematics educators.* Mathematics education faculty members, whether within colleges of education, mathematics departments, or other academic units, are also an important audience for this report. Typically, they are responsible for the pedagogical education of mathematics teachers (e.g., teaching methods courses), organizing field experiences for prospective teachers, and for providing overall leadership for the institution's mathematics teacher preparation program. Outside of academe, a variety of people are engaged in professional development for teachers of mathematics, including state, regional, and school-district mathematics specialists. The term "mathematics educators" will include this audience.

Policy-makers. The report is also intended to inform educational adminis-trators and policy-makers at the national, state, school-district, and colle-giate levels as they work to provide PreK–12 students with a strong math-ematical preparation for the increasingly quantitative workplace. Teach-ers' knowledge of mathematics is central to this effort, thus, institutions of higher education have a key role to play in teachers' professional devel-opment as well as their preparation.

Teachers' learning of mathematics is supported—or hindered—by in-stitutional policies. These include national accreditation requirements, state certification requirements, and the ways in which they are reflected in teacher preparation programs. State and district supervisors make choices in provision and funding of professional development. At the school level, scheduling and policy affect the types of learning experiences available to teachers. Thus, policy-makers play important roles in the mathematical education of teachers.

Terminology. To avoid confusion, the report uses the following terminology:

Student refers to a child or adolescent in a PreK–12 classroom.

Teacher refers to an instructor in a PreK–12 classroom but may also refer to a prospective PreK–12 teacher in a college mathematics course ("prospective teacher" or "pre-service teacher" is also used in the latter case).

Instructor refers to an instructor of prospective or practicing teachers. Because this report concerns the roles of mathematicians in teacher edu-cation, "instructor" will usually refer to a mathematician.

Acknowledgements. The work of preparing the MET II report was made possible by a grant from Math for America.

The content and exposition of this report has benefited from extensive and thoughtful criticism of an earlier draft from teachers, mathematicians, and mathe-matics educators.

That earlier draft drew on comments and suggestions made by participants at the 2010 CBMS Forum on Content-Based Professional Development and the 2011 CBMS Forum on Teaching Teachers in the Era of the Common Core. These were made possible by support from the Brookhill Foundation and the National Science Foundation.

CHAPTER 1

School Mathematics and Teachers' Mathematics

A critical pillar of a strong PreK–12 education is a well-qualified teacher in every classroom. This report offers recommendations for the mathematical preparation and professional development of such teachers.

A second pillar is a challenging, world-class curriculum. In mathematics, the substance for this pillar is supplied by the Common Core State Standards (CCSS). These standards are created from progressions: sequences of topics and performances designed to respect the structure of mathematics and cognitive aspects of learning mathematics. This report focuses on teachers' knowledge of the mathematical aspects of these progressions: the sequences of topics and the mathematical structures that underlie these sequences.[1]

The CCSS also include standards for mathematical practice.[2] Their formulation was influenced by the National Council of Teachers of Mathematics process standards, the elements of mathematical proficiency described in the National Research Council report *Adding It Up*, and the discussions of the Park City Mathematics Standards Study Group.[3] Like their students, teachers need to have the varieties of expertise described in these standards—monitoring their own progress as they solve problems, attending to precision, constructing viable arguments, seeking and using mathematical structure, and making strategic use of appropriate tools, e.g., notations, diagrams, graphs, or procedures (whether implemented by hand or electronically). These abilities are supported by the mathematical "habits of mind" described in the original MET report.

At every grade level—elementary, middle, and high school—there is important mathematics that is both intellectually demanding to learn and widely used, such as reasoning strategies that rely on base-ten algorithms in elementary school; ratio, proportion, and exploratory statistics in middle school; algebra, geometry, and data analysis in high school. Teachers need to have more than a student's understanding of the mathematics in these grades. To support curricular coherence, teachers need to know how the mathematics they teach is connected with that of prior and later grades.[4] For example, an elementary teacher needs to know how the associative, commutative, and distributive properties are used together with place value in algorithms for addition and multiplication of whole numbers, and the

---

[1]An overview of the CCSS structure appears as Appendix B of this report.

[2]The full text of these standards appears as Appendix C.

[3]Between 2004 and 2008, the Park City Mathematics Study Group (a group of research mathematicians) conducted discussions of school mathematics, including extended discussions with NCTM representatives. *Principles and Standards* and *Adding It Up* (published in 2000 and 2001) summarize findings from previous decades of research in mathematics education.

[4]Such connections are outlined in the Progressions for the CCSS (see the web resources for this report).

significance of these algorithms for decimal arithmetic in later grades. In the middle grades, a teacher needs to know how to build on this foundation; for instance, how to help students to extend these algorithms correctly to decimals and to use the distributive and other properties when adding and subtracting linear expressions. A high school teacher builds on the same ideas in teaching students about calculations with polynomials and other symbolic expressions.

Moreover, to appropriately create, select, or modify tasks, teachers need to understand the mathematical consequences of different choices of numbers, manipulative tools, or problem contexts.[5] They need to recognize the need for definitions (e.g., "What is a fraction?," "What does it mean to add two fractions?") and their consequences ("How do we know that the sum is unambiguously determined?"). Concepts may be defined differently in different resources being used, whether text-based or online (e.g., a trapezoid has at least one pair of parallel sides versus exactly one pair), and have different consequences (e.g., parallelograms are trapezoids—or not). Different assumptions also have different consequences. For example, in discussing properties of numbers ("Does 'number' mean 'whole number' or 'fraction'?"), in geometry ("Does this depend on the parallel postulate?"), or in modeling ("Is the flow uniform or not?").

Software, manipulatives, and many other tools exist to support teaching and learning. In order to use these strategically in teaching, and to help students use them strategically in doing mathematics, teachers need to understand the mathematical aspects of these tools and their uses. Teachers need the ability to find flaws in students' arguments, and to help their students understand the nature of the errors. Teachers need to know the structures that occur in school mathematics, and to help students perceive them.

The technical knowledge inherent in these examples implies that the profession of mathematics teaching requires a high level of expertise.[6] International and domestic studies suggest that an important factor in student success is a highly skilled teaching corps,[7] and that teachers' expertise is developed or hindered by institutional arrangements and professional practices.[8] Widespread expertise is aided by high standards for entry into the profession, and continual improvement of mathematical knowledge and teaching skills. Continual improvement can be promoted by regular interactions among teachers, mathematicians, and mathematics education faculty in creating and analyzing lessons, textbooks, and curriculum

---

[5]Examples are given by Ma, *Knowing and Teaching Elementary Mathematics*, Erlbaum, 1999: changes in number, p. 74; change in manipulative and problem context, p. 5.

[6]For a summary (p. 400) and further examples of teaching tasks, see Ball et al., "Content Knowledge for Teaching," *Journal of Teacher Education*, 2008; also Senk et al., "Knowledge of Future Primary Teachers for Teaching Mathematics: An International Comparative Study," *ZDM*, 2012, p. 310.

[7]See, e.g., the findings of the Teacher Education and Development Study in Mathematics (TEDS-M).

[8]These are intertwined and occur on a variety of levels. For example, the institutional arrangement of having teachers share a room affords the professional practice of discussing mathematics. An institutionalized career hierarchy based on teaching shapes the professional activities of Chinese master teachers and "super rank" teachers described in *The Teacher Development Continuum in the United States and China*, National Academies Press, 2010. In Japan, institutional arrangements afford the practice of "lesson study," allowing teachers to communicate with other teachers in their school or district, and with policy-makers (see Lewis, *Lesson Study*, Research for Better Schools, 2002, pp. 20–22).

documents; and examining the underlying mathematics.[9] To support the spread of expertise in PreK–12 mathematics teaching, the mathematical education of teachers should become a central concern of the mathematics community. In particular, the mathematical education of teachers will need to become a central concern of more mathematicians and collegiate mathematics departments.

Current efforts to improve PreK–12 mathematics education in the United States recognize that school systems, communities, families, and teachers, as well as students themselves, all share responsibility for high-quality mathematics learning.[10] In a similar fashion, high-quality mathematical education of teachers is the responsibility of institutions of higher education, professional societies, accrediting organizations, and school districts, as well as PreK–12 teachers themselves. Their collective goal needs to be continual improvement in the preparation and further education of mathematics teachers.

This report describes the mathematical knowledge that teachers at different levels need. It puts special emphasis on professional development, because mathematical knowledge for teaching can and should continue to grow throughout a teacher's career. The report discusses the kinds of experiences that can create, extend, and deepen knowledge at each stage of a teacher's career:

    i. opportunities for beginning teachers;

    ii. increasing expertise for teachers with several years experience;

    iii. enhancing the skills of very experienced teachers.

Collegiate mathematics faculty members have vital roles to play in these experiences, and this report describes how they can contribute in productive ways.

Professional development should include self-directed study as well as activities that involve school-district mathematics supervisors and faculty in mathematics education and mathematics. To assist mathematics faculty with little experience in offering professional development opportunities for teachers, this report draws on the experiences of a range of professional development programs funded by the National Science Foundation and United States Department of Education's Math Science Partnerships, and other foundation- and public-sector-based initiatives. Interested readers are invited to learn more about these programs and contact program leaders for assistance in adopting and adapting the programs to their locations (see the web resources associated with this report).

Each different level of teacher education presents particular challenges for the education of mathematics teachers. Perhaps the most publicized challenges involve

---

[9]Chapter 2 discusses this claim further, but note the findings of *Effects of Teacher Professional Development on Gains in Student Achievement*, Council of Chief State School Officers, 2009. Most successful professional development programs continued for 6 months or more, and the mean contact time with teachers was 91 hours.

[10]For example, the Mathematics Common Core Coalition (comprised of professional societies and assessment consortia) addresses educators, teachers, teacher leaders, supervisors, administrators, governors and their staffs, other policy-makers, and parents.

the education of elementary teachers. Like many undergraduates,[11] future elementary teachers may enter college with only a superficial knowledge of K–12 mathematics, including the mathematics that they intend to teach. For example, they may not know rationales for computations with fractions or the role of place value in base-ten algorithms, and may not have the opportunity to learn them as undergraduates.[12] Moreover, much that is useful to teachers is known about teaching–learning paths for early mathematics,[13] but, too often mathematicians who are new to this area lack the knowledge or resources to help future teachers develop an understanding of these paths and their mathematical stepping-stones.[14] After elementary teachers begin teaching, it is rare for them to have any sustained professional development centered on mathematics.[15] This report's recommendations for elementary teachers call for comprehensive professional development programs in mathematics coupled with more in-depth pre-service study of school mathematics. To do this, the recommended number of semester-hours of mathematics courses specifically designed for teachers is raised to 12 from the original MET Report's 9.

Far too frequently, middle grades teachers have the same preparation as elementary generalists.[16] This must stop. This report repeats the original MET Report's recommendation that grades 5–8 mathematics be taught by teachers who specialize in this subject and raises the recommended number of semester-hours in

---

[11]The CBMS surveys (conducted every five years) consistently document large proportions of undergraduates enrolled in remedial mathematics courses (see, e.g., Table S.2 of the 2005 report).

[12]The 2005 CBMS survey suggests that many mathematics departments do not have courses especially designed for elementary teachers (see Table SP.6). In 2010, Masingila et al. surveyed 1,926 U.S. higher education institutions that prepared elementary teachers. Of those who responded (43%), about half (54%) reported that requirements included two mathematics courses designed for teachers. See "Who Teaches Mathematics Content Courses for Prospective Elementary Teachers in the United States? Results of a National Survey," *Journal of Mathematics Teacher Education*, 2012, Table 2. A more detailed picture for three states is presented by McCrory & Cannata, "Mathematics Classes for Future Elementary Teachers: Data from Mathematics Departments," *Notices of the American Mathematical Society*, 2011.

[13]Chapter 2 gives an overview of teaching–learning paths.

[14]In Masingila et al.'s survey less than half of respondents reported giving training or support to instructors of mathematics courses for elementary teachers.

[15]For example, when surveyed in 2000, 86% of K–4 teachers reported studying mathematics for less than 35 hours over a period of three years, an average of less than 12 hours per year. See Horizon Research's 2000 National Survey of Science and Mathematics Education. More recent studies show large increases in elementary student mathematics achievement when their teachers receive content-based professional development. Student scores of teachers who do not receive such professional development do not show these gains (see the sections on curriculum-specific professional development in Chapter 2 and on mathematics specialists in Chapter 4). Thus, unsatisfactory student performance may suggest a greater need for content-based professional development.

[16]The Association for Middle Level Education (AMLE) position statement notes, "in some states, virtually anyone with any kind of degree or licensure is permitted to teach young adolescents." According to the AMLE web site, 28 states and the District of Columbia offer separate licenses for middle grades generalists. Separate licenses, however, do not necessarily imply the existence of separate preparation programs or different mathematics requirements. The 2005 CBMS survey found that 56% of mathematics departments at four-year institutions had the same mathematics requirements for K–8 certification in early and later grades (see Table SP.5). See also the discussion of opportunity to learn for U.S. prospective lower secondary teachers in Tatto & Senk, "The Mathematics Education of Future Primary and Secondary Teachers: Methods and Findings from the Teacher Education and Development Study in Mathematics," *Journal of Mathematics Teacher Education*, 2011, p. 127.

mathematics to 24. All states need to institute certification programs for middle grades mathematics teachers.

Because many practicing middle grades mathematics teachers received certification by meeting expectations that were more appropriate for elementary teachers, opportunities for content-based professional development are needed that address their situation. This need is even more critical in the context of the increased expectations indicated by the CCSS.

Although high school mathematics teachers frequently major in mathematics, too often the mathematics courses they take emphasize preparation for graduate study or careers in business rather than advanced perspectives on the mathematics that is taught in high school. This report offers suggestions for rethinking courses in the mathematics major in order to provide opportunities for future teachers to learn the mathematics they need to know to be well-prepared beginning teachers who will continue to learn new mathematical content and deepen their understanding of familiar topics. As stated in MET I, "college mathematics courses should be designed to prepare prospective teachers for the life-long learning of mathematics, rather than to teach them all they will need to know." This viewpoint is especially important in the context of the greater sophistication and breadth of the mathematical expectations for high school students described by the CCSS.

CHAPTER 2

The Mathematical Education of Teachers: Traditions, Research, Current Context

This report focuses on the mathematical education of teachers, asking more mathematics departments and more mathematicians to assign high priority to teacher preparation, content-based professional development, partnerships with mathematics educators, and increased participation in the mathematics education community. To appreciate the need for changing some current priorities and practices in teacher education, it is important to understand what they are, and the traditions of school mathematics that shaped them, and still shape prospective and practicing teachers. Thus, this chapter briefly reviews traditions of teacher education and school mathematics. It is also helpful to review what is known about the mathematical knowledge needed for teaching. Thus, this chapter gives an overview of current research on teacher knowledge, and discusses it in light of the Common Core State Standards and other aspects of the current educational context.

Traditions, Beliefs, and Practices

Mathematicians' roles in teacher education. As stewards of their discipline, mathematicians have a long tradition of concerning themselves with school mathematics and its teachers. In the eighteenth century, Leonard Euler wrote an arithmetic textbook as did Augustus de Morgan a century later.[1] Felix Klein's work with high school teachers gave us the notion of "elementary mathematics from an advanced standpoint"—understanding the mathematical foundations of school mathematics. Klein was a founder of what is now the International Commission on Mathematical Instruction. Since its inception in 1908 as part of the International Mathematics Union, its presidents have included Jacques Hadamard, Marshall Stone, and other distinguished mathematicians.[2]

In the United States, as in many other countries, mathematicians' involvement in teacher preparation increased as nineteenth-century normal schools became twentieth-century colleges and universities. In 1893, the Committee of Ten, composed of presidents of Harvard and other leading universities, led the creation of influential school curriculum guidelines. Among the writers were Simon Newcomb and Henry Fine, both future presidents of the American Mathematical Society.

However, for a variety of reasons, both internal and external to the U.S. mathematics community, concern for school mathematics and its teachers did not retain

---

[1]These were: *Einleitung zur Rechen-Kunst* (*Introduction to the Art of Reckoning*), St Petersburg (vol. 1, 1738, vol. 2, 1740); *The Elements of Arithmetic*, London, 1830.

[2]Hodgson points out that "one could even see the ICMI as having been formed on the very assumption that university mathematicians should have an influence on school mathematics." See *The Teaching and Learning of Mathematics at University Level*, Kluwer, 2001, p. 503.

similar prominence among mathematicians during much of the twentieth century.[3] Although there have been notable counterexamples,[4] teacher education and school mathematics have often been peripheral concerns for mathematicians and mathematics departments. This situation is consistent with existing policies and practices, inside and outside of mathematics departments. Departmental support and professional development for mathematicians involved with teacher education is often sparse.[5] In the past, professional development centered on mathematics for PreK–12 teachers has been infrequent, both in general and as an activity of collegiate mathematics departments. Over the past decade, this situation has begun to change. An aim of this report is to facilitate further change.

Beliefs about mathematics and their influences on learning. As mathematicians' involvement with school mathematics decreased, the U.S. educational system expanded. Beliefs evolved—or were maintained—that shape the context of education today. Among these were students' beliefs about mathematics.

In the 1980s, education researchers began to document unmathematical beliefs among K–12 students. The statements below summarize observations of high school geometry classes where homework sets consisted of 18 to 45 problems. (Note that the first statement is counter to the first Common Core Standard for Mathematical Practice: "Make sense of problems and persevere in solving them.")

- Students who have understood the mathematics they have studied will be able to solve any assigned problem in five minutes or less.

- Ordinary students cannot expect to understand mathematics: they expect simply to memorize it and apply what they have learned mechanically and without understanding.[6]

---

[3]Murray discusses the polarization of teaching and research within the U.S. mathematical community in *Women Becoming Mathematicians: Creating a Professional Identity in Post–World War II America*, MIT Press, 2000, pp. 6–10. For examples of U.S. mathematician involvement (e.g., the founding of the International Commission on the Teaching of Mathematics (later ICMI) at the International Congress of Mathematicians) and social context of its diminution, see Donoghue, "The Emergence of a Profession: Mathematics Education in the United States, 1890–1920," in *A History of School Mathematics*, vol. 1, NCTM, 2003. Changes in twentieth-century psychology research were also a factor, see Roberts, "E. H. Moore's Early Twentieth-Century Program for Reform in Mathematics Education," *American Mathematical Monthly*, 2001.

[4]*Teaching Teachers Mathematics* (Mathematical Sciences Research Institute, 2009) gives an overview of past and recent counterexamples.

[5]In 2010, Masingila et al. surveyed 1,926 U.S. higher education institutions that prepared elementary teachers. Of those who responded (43%), less than half reported giving training or support for instructors of mathematics courses for elementary teachers. However, the authors write that "there appears to be interest in training and support as a number of survey respondents contacted us to ask where they could find resources for teaching these courses." See "Who Teaches Mathematics Content Courses for Prospective Elementary Teachers in the United States? Results of a National Survey," *Journal of Mathematics Teacher Education*, 2012.

[6]Quoted from Schoenfeld, "Learning to Think Mathematically" in *Handbook for Research on Mathematics Teaching and Learning*, 1992, p. 359. Note that these beliefs may not be explicitly stated as survey or interview responses, but displayed as classroom behaviors, e.g., giving up if a problem is not quickly solved. This discussion is not meant to exclude the possibility of exceptional mathematical talent, but focuses on the idea that K–12 mathematics can be learned in its absence.

Although education researchers have identified these and other unproductive beliefs held by K–12 students, experience and other lines of research suggest that adults may hold similar beliefs about the existence of people with "math minds" or the existence of a "math gene."[7]

Recent psychological research suggests that such beliefs influence teaching and learning. This line of research has identified two distinct views. The "fixed mind-set" or "entity view of intelligence" considers cognitive abilities to be fixed from birth or unchangeable. In contrast, the "growth mind-set" or "incrementalist view" sees cognitive abilities as expandable.[8] International comparisons suggest that different views are associated with differences in achievement, and research within the U.S. has documented such associations. Students who entered seventh grade with a growth mind-set earned better grades over the next two years than peers who entered with a fixed mind-set and the same scores on mathematics tests. Classroom studies have shown that it is possible to change students' views from a fixed mind-set to a growth mind-set in ways that encourage them to persevere in learning mathematics and improve achievement test scores as well as grades.[9] Studies like these suggest that teaching practices are an important factor in reinforcing or changing students' beliefs.

Practices in teaching mathematics and their influence on learning. Unproductive beliefs about mathematics were identified in the late twentieth century, but historical research suggests that they may have been fostered by early schooling practices. Among these were pedagogical approaches. The "rule method" (memorize a rule, then practice using it) was the sole approach used in U.S. arithmetic textbooks from colonial times until the 1820s.[10] Between 1920 and 1930, pedagogy based on the work of the psychologist Edward Thorndike again emphasized memorization, e.g., memorization of arithmetic "facts" with no attempt to encourage children to notice how two facts might be related. Thus, $3 + 1 = 4$ was not connected to $1 + 3 = 4$, missing an opportunity to begin developing an understanding of the commutative law as well as the mathematical practice of seeking structure (see Appendix C). These pedagogical ideas were revived in the "back to basics" era of the 1980s and are sometimes still used, despite the existence of very different approaches that are currently used.[11]

---

[7]Stevenson and Stigler documented similar beliefs among U.S. first and fifth graders, and their mothers, but found that their Japanese and Chinese counterparts focused more on effort rather than ability. See Chapter 5 of *The Learning Gap*, Simon & Schuster, 1992. See also *Data Compendium for the NAEP 1992 Mathematics Assessment for the Nation and the States*, National Center for Educational Statistics, 1993.

[8]Note that such beliefs may vary according to domain, e.g., one may believe in a "math gene," but favor continued practice in order to improve sports performance.

[9]For a brief overview of research in this area, including classroom studies, see Dweck, "Mind-sets and Equitable Education," *Principal Leadership*, 2010. For a review of research and recommendations for classroom practice, see *Encouraging Girls in Math and Science* (IES Practice Guide, NCER 2007-2003), Institute of Educational Sciences, 2007, pp. 11–13.

[10]See Michalowicz & Howard, "An Analysis of Mathematics Texts from the Nineteenth Century" in *A History of School Mathematics*, vol. 1, NCTM, 2003, especially pp. 82–83.

[11]Lambdin & Walcott, "Changes through the Years: Connections between Psychological Learning Theories and the School Mathematics Curriculum," *The Learning of Mathematics*, 69th Yearbook, NCTM, 2007. For discussion of current practices, see Ma, "Three Approaches to One-Place Addition and Subtraction: Counting Strategies, Memorized Facts, and Thinking Tools," unpublished.

Other beliefs may have maintained fragile understanding of mathematics for teachers and their students, reinforcing teachers' reliance on approaches that focused on memorizing and following rules. One was the belief that elementary teachers learned all the mathematics that they needed to know during their own schooling. Such beliefs are reflected in the policies and practices noted in Chapter 1: few or no mathematics requirements for K–8 teacher preparation and certification; and infrequent professional development centered on mathematics.

In addition to identifying counterproductive beliefs about learning mathematics, mathematics education researchers have identified associated beliefs about the roles of teachers and students in mathematics classrooms:

- Doing mathematics means following the rules laid down by the teacher.

- Knowing mathematics means remembering and applying the correct rule when the teacher asks a question.

- Mathematical truth is determined when the answer is ratified by the teacher.[12]

Systematic studies of U.S. classrooms are not abundant, but their findings and those of student surveys are consistent with these descriptions of classroom expectations.[13]

Consistent with traditions for classroom behavior, videotape analyses have found far fewer occurrences of deductive reasoning in U.S. mathematics classrooms than in classrooms from countries whose students score well on international tests.[14] Moreover, studies of U.S. textbooks and curriculum documents suggest that they have often been constructed in ways that do not readily afford deductive reasoning. Such curriculum studies note imprecise, nonexistent, or contradictory definitions, or more global issues such as repetition of topics, suggesting disconnected treatments of topics with similar underlying structures (e.g., base-ten notation for whole numbers and for decimals).[15]

Summary. These traditions in U.S. school mathematics suggest that undergraduates (including prospective teachers) who have been educated in the U.S. may have well-established beliefs about mathematics and expectations for mathematics instruction that are antithetical to those of their mathematician instructors. As stated in MET I:

---

[12]This is a slight reformulation of Lampert, 1990 as quoted by Schoenfeld, "Learning to Think Mathematically" in *Handbook for Research on Mathematics Teaching and Learning,* 1992, p. 359. The surrounding text discusses research on school experiences that shape such beliefs.

[13]For example, see Hiebert et al.'s study of eighth grade classrooms, *Teaching Mathematics in Seven Countries: Results from the TIMSS 1999 Video Study,* U.S. Department of Education, 2003.

[14]See analyses of data from the TIMSS video studies of 1999 (Hiebert et al., pp. 73–75) and of 1995 (Manaster, *American Mathematical Monthly,* 1998).

[15]Schmidt and Houang analyzed the content and sequencing of topics in grades 1–8 in the U.S. and other countries. See "Lack of Focus in the Mathematics Curriculum," in *Lessons Learned,* Brookings Institution Press, 2007, p. 66. Examples of treatments of fractions and negative numbers that do not afford deductive reasoning are given by Wu in "Phoenix Rising," *American Educator,* 2011.

> For many prospective teachers, learning mathematics has meant *only* learning its procedures and, they may, in fact, have been rewarded with high grades in mathematics for their fluency in using procedures. (emphasis added)

The traditions and findings described here suggest that doing mathematics in ways consistent with mathematical practice is likely to be a new, and perhaps, alien experience for many teachers. However, such experiences are necessary for teachers if their students are to achieve the Common Core State Standards for Mathematical Practice.

Although this situation may look grim, it is not intractable. Collaborations between mathematicians and mathematics educators in teacher education have made remarkable progress in developing ways to address teachers' unmathematical beliefs and practices as well as gaps in their mathematical knowledge.[16] As evidenced by outcomes from the Math Science Partnerships and research on professional development, teachers can acquire mathematical practices from carefully designed experiences of doing mathematics.[17] This suggests that doing mathematics in ways consistent with the Common Core State Standards for Mathematical Practice is an important element in the mathematical education of teachers.

Teacher Effectiveness and Mathematical Knowledge

"Teacher effectiveness" is generally construed as the effect that a teacher has on her or his students' learning. Research on teacher effectiveness often examines relationships between teacher knowledge and student achievement. In these studies, students' achievement is generally measured by standardized tests,[18] but their teachers' knowledge has been measured in quite different ways.

Mathematics courses and certification. For at least 50 years, studies of teacher effectiveness have often focused on teacher preparation, and mainly on high school and middle grades teachers. Certification status has been a popular measure. The existing evidence suggests that certification in mathematics is desirable for high school and middle grades teachers. Another measure has been the number and type of mathematics courses taken. In general, studies of high school and middle grades teachers report that more mathematics courses are associated with better performance by their students. However, these effects are small, sometimes inconsistent, and do not indicate the type of knowledge used in teaching.[19] Moreover,

---

[16]For example, middle grades and high school teachers who participated in an MSP based on an immersion approach (involving intensive sessions of doing mathematics) reported changes in beliefs that affected their teaching, e.g., communicating that it is "OK" to struggle. See *Focus on Mathematics Summative Evaluation Report 2009*, p. 73. Gains in student test scores are shown on p. 93 (high school) and p. 96 (middle grades).

[17]For a snapshot from one such collaboration, see *Teaching Teachers Mathematics*, Mathematical Sciences Research Institute, 2009, p. 34; for descriptions of three Math Science Partnerships, see pp. 32–41.

[18]Test quality can be a major limitation for this measure. An analysis of state mathematics tests found low levels of cognitive demand, e.g., questions that asked for recall or performance of simple algorithms, rather than complex reasoning over an extended period. See Hyde et al., "Gender Similarities Characterize Math Performance," *Science*, 2008, pp. 494–495.

[19]See *Preparing Teachers: Building Evidence for Sound Policy*, National Research Council, 2010, p. 112. See also, Telese, "Middle School Mathematics Teachers' Professional Development

certification or undergraduate course-taking are quite imprecise measures, due to variability in certification requirements and undergraduate instruction.

Mathematical knowledge for teaching. A different line of research has begun to offer evidence that particular forms of mathematical knowledge are important in teaching. In the 1980s, scholars began to investigate "knowledge for teaching," criticizing earlier research on effectiveness for ignoring the subject matter and its transformation into the content of instruction.[20] Initially, this line of research analyzed the actions of teachers in classrooms or outcomes of interviews with teachers, rather than survey data and test scores. The focus was on identifying kinds of knowledge relevant for *teaching* mathematics, rather than mathematical knowledge in general. For example, prospective teachers were asked to respond to classroom scenarios, such as a question about why division by 0 is undefined. Responses indicated that even mathematics majors were not always able to answer in a satisfactory way.[21]

As noted in MET I, such interviews with teachers awakened many mathematicians to the special nature of mathematics for teaching and its implications for the education of teachers. Since that time, this line of research has continued toward developing tests of mathematical knowledge for teaching. Third-grade teachers' scores on one such test (Learning Mathematics for Teaching) were better predictors of their students' achievement than measures such as average time spent in mathematics instruction, years of experience, and certification status.[22]

Curriculum-specific professional development. A second line of recent research has focused on studying relationships between teachers' professional development experiences and their students' performance on mathematics tests. A 1998 study of professional development in California found that attending workshops that were mathematics- and curriculum-specific (e.g., as opposed to learning to use manipulatives or to improve classroom management) was associated with better student performance on mathematics tests.[23] A 2009 meta-analysis of professional development studies found that those in which teachers focused, for a sustained period, on examining mathematics underlying the curriculum and how to teach it were associated with improved student achievement.[24] Similarly, a project in which a research-based "toolkit" on fractions was supplied to treatment groups of U.S. elementary teachers to use in lesson study found that groups who used the toolkit

---

and Student Achievement," *Journal of Educational Research*, 2012. Telese's measure of student achievement was the Grade 8 National Assessment of Educational Progress, which includes items with a high level of cognitive demand. It found number of mathematics courses to be a strong predictor, but like many such studies, it did not have an experimental or quasi-experimental design.

[20]Shulman, "Those Who Understand: Knowledge Growth in Teaching," *Educational Researcher*, 1986.

[21]On average, the prospective secondary teachers had taken over 9 college-level mathematics courses. Ball, "Prospective Elementary and Secondary Teachers' Understanding of Division," *Journal for Research in Mathematics Education*, 1990.

[22]Hill et al., "Effects of Teachers' Mathematical Knowledge for Teaching on Student Achievement," *American Educational Research Journal*, 2005.

[23]Cohen & Hill, "Instructional Policy and Classroom Performance: The Mathematics Reform in California," *Teachers College Record*, 2000.

[24]Blank & Atlas, *Effects of Teacher Professional Development on Gains in Student Achievement*, Council of Chief State School Officers, 2009.

were associated with significantly greater student achievement than those of control groups.[25]

Teaching–learning paths. A third line of research on teacher effectiveness focuses on learning trajectories—sequences of student behaviors indicating different levels of thinking with instructional tasks that lead to development of a mathematical ability. Related ways to focus instruction are described as teaching–learning paths, "learning lines," and learning progressions. These notions, together with examples of paths from U.S. research and curriculum materials from other countries, informed the development of the CCSS.

An example from China may help to illustrate the general nature of these U.S. notions. Chinese teachers describe a sequence of problems together with concepts and skills that lead students to be able to compute whole-number subtraction problems with regrouping (e.g., $104 - 68$), and to understand the rationale for their computations. Each part of the sequence involves a new kind of problem, a new idea, and a new skill.

Minuends between 10 and 20, e.g., $15 - 7$, $16 - 8$	New concept and skill of decomposing a ten.
Minuends between 19 and 100, e.g., $53 - 25$, $72 - 48$	New concept and skill of splitting off a ten, followed by decomposing a ten.
Minuends with three or more digits.	New concept and skill of successive decomposition.[26]

In the U.S., randomized studies of preschool classrooms have shown large student gains for a curriculum based on learning trajectories that included sustained and specific professional development for teachers.[27] Studies of elementary grades have focused on assessment tasks, rather than entire curricula. But, like the curriculum for the preschool classrooms, these tasks outline a learning path that goes step by step, helping students incrementally increase their understanding, as they move toward a mathematical goal. They also create a teaching path, helping teachers perceive the elements of a given concept or skill, and mathematical stepping-stones in their development.[28]

Large-scale studies that examine connections between student achievement in earlier and later grades suggest that improved mathematics instruction in preschool and elementary grades has a large payoff in later achievement, not only for mathematics in later grades (including high school), but for reading.[29] Such studies

---

[25]Perry & Lewis, *Improving the Mathematical Content Base of Lesson Study: Summary of Results*, 2011.

[26]Example from Ma, *Knowing and Teaching Elementary Mathematics*, Erlbaum, 1999, p. 15. Similar examples occur in other East Asian countries. Lewis et al. describe how Japanese teacher's manuals may support teachers' perceptions of paths in "Using Japanese Curriculum Materials to Support Lesson Study Outside Japan: Toward Coherent Curriculum," *Educational Studies in Japan: International Yearbook*, 2011.

[27]Sarama & Clements, *Early Childhood Mathematics Education Research*, Routledge, 2009, pp. 352–363.

[28]See special issue on learning trajectories, *Mathematical Thinking and Learning*, 2004.

[29]See Duncan et al., "School Readiness and Later Achievement," *Developmental Psychology*, 2007; Claessens et al., "Kindergarten Skills and Fifth-grade Achievement: Evidence from the ECLS-K," *Economics of Education Review*, 2009; Siegler et al., "Early Predictors of High

reiterate the importance of mathematics in preparation and professional development for early childhood and elementary teachers.

Summary. Studies of teacher effectiveness suggest that mathematics course-taking and certification are desirable for middle grades and high school teachers, but are inconclusive about the nature of the mathematical knowledge that teachers need. However, the existing evidence suggests that teacher preparation and professional development should be tailored to the work of teaching. The National Research Council study *Preparing Teachers* concludes:

> Current research and professional consensus correspond in suggesting that all mathematics teachers ... rely on: mathematical knowledge for teaching, that is, knowledge not just of the content they are responsible for teaching, but also of the broader mathematical context for that knowledge and the connections between the material they teach and other important mathematics content.[30]

Within the U.S., such knowledge is not currently well developed in the profession of mathematics teaching. Mathematicians are among those necessary for its development.

For PreK–8 teachers, adequate preparation includes more mathematics than often thought. Moreover, studies connecting teachers' understanding of teaching–learning paths and student achievement show how the organization of curriculum together with attention to teacher knowledge can work together to improve students' learning. A necessary first step for teachers is to understand the mathematics in these paths,[31] thus mathematicians' participation in their education is essential.

Current Context

Since MET I was published in 2001, there have been significant changes in teacher education: outside mathematics departments with respect to the teaching workforce and educational policy; within mathematics departments with respect to courses for teachers and faculty involvement in K–12 education.

Demographic changes have occurred for the teaching workforce as a whole. Analyses of nationally representative survey data find that between 1988 and 2008, the age distribution for teachers shifted from a unimodal distribution with a peak at age 41 to a bimodal distribution with peaks at ages 26 and 55. Some of these changes appear to be due to increases in the numbers of teachers for special education, elementary enrichment, science, and mathematics.[32]

In 2000, approximately 22% of secondary schools reported serious difficulties in filling teaching positions for mathematics. This dropped to about 18% in 2008. Such staffing difficulties tended to occur at high-poverty, high-minority public schools in both urban and rural areas. Over half of the teachers who left these

---

School Mathematics Achievement," *Psychological Science*, 2012. These studies examined large longitudinal data sets from the U.S. and other countries.

[30] *Preparing Teachers: Building Sound Evidence for Sound Policy*, National Research Council, 2010, pp. 114–115.

[31] This is made explicit for early childhood educators in *Mathematics Learning in Early Childhood*, National Research Council, 2009, pp. 3–4.

[32] Ingersoll & Merrill, "Who's Teaching Our Children?," *Educational Leadership*, 2010.

schools reported dissatisfaction or the intention to pursue another or better job.[33] Analysis of 2004 and 2005 data found differences in rates of teacher attrition at schools in the same district. Also, mathematics teachers who moved from one teaching job to another were most likely to move to schools with similar enrollments of poor and minority students. This suggests that attrition is not simply a matter of school demographics, but of school organization. An organizational factor of particular relevance to the MET II report is provision of content-based professional development. Mathematics teachers who received it and perceived it as useful had substantially lower odds of turnover.[34]

About 40% of practicing teachers have been prepared via an alternative pathway, that is, outside of a traditional teacher education program. Like standard programs, these alternative pathways vary widely.[35] Such differences can affect the rest of a teacher's career. Analyses of recent survey data find that in the first year of teaching, teachers with a mathematics baccalaureate, but little or no pedagogical preparation, left teaching at twice the rate of those with the same degree, but more comprehensive pedagogical preparation.[36]

A new accreditation organization with significantly different standards for teacher preparation is coming into existence. The Council for the Accreditation of Educator Preparation (CAEP) will require that the mathematical preparation of teachers address the CCSS.[37] In the past, accreditation requirements for mathematics have often been met by reporting results on tests such as the Praxis or course grades for appropriate courses, although other options were available. The new requirements for mathematics courses will be similar in nature to the current, more detailed, accreditation requirements for methods courses. The standards for these courses have changed to include standards for mathematical practice and to reflect the content of the CCSS.

Requirements for professional development have been changing. By 2008, all 50 states had specified professional development requirements for teachers. The majority of these require 6 semester-hours of professional development over approximately 5 years. Twenty-four of these states have a policy specifying that professional development be aligned with state content standards.[38]

More mathematics departments have designed courses especially for K–8 teachers or have designated special sections of courses for these teachers.[39] In some

---

[33]Ingersoll & Perda, "Is the Supply of Mathematics and Science Teachers Sufficient?," *American Educational Research Journal*, 2010.

[34]Ingersoll & May, "The Magnitude, Destinations, and Determinants of Mathematics and Science Teacher Turnover," Consortium for Policy Research in Education, 2010, pp. 44, 46.

[35]*Preparing Teachers: Building Sound Evidence for Sound Policy*, National Research Council, 2010, pp. 34–39.

[36]Ingersoll & Merrill, "Retaining Teachers: How Preparation Matters," *Educational Leadership*, 2012. See also Darling-Hammond, *Solving the Dilemmas of Teacher Supply, Demand, and Standards*, National Commission on Teaching and America's Future, 2000, pp. 17–19; *Tenth Anniversary Report*, UTeach, 2010, p. 16.

[37]CAEP was formed by the merger of the National Council for the Accreditation of Teacher Education (NCATE) and the Teacher Education Accreditation Council (TEAC). Two of the MET II writers are engaged in the development of the CAEP standards.

[38]*Key State Education Policies on PK–12 Education: 2008*, Council of Chief State School Officers, p. 22.

[39]CBMS 2005 Survey, Table SP.3.

departments, policies for faculty have changed to facilitate their involvement in activities for increasing K–12 student achievement.[40]

Collectively, the mathematics community now has substantial experience in developing partnerships that allow teachers to achieve the goals for teacher preparation and professional development described in this report and others.[41] Collaboration with others in mathematics education has allowed mathematicians to have a major impact on professional development within states.[42] Partnerships that began in the 1990s have expanded in scope or have been duplicated at multiple locations. Mathematicians have expanded their involvement in mathematics education, forming partnerships with mathematics education researchers, education officials, and teachers in new kinds of programs. Through these experiences, concerned mathematicians gained greater expertise and awareness about the challenges to improving mathematical learning in the schools, and within states. More information about these and other relevant efforts is on the web page associated with this report. Because the CCSS have been adopted by most states, many of these projects will be able to share details and specifics about students' and teachers' learning of mathematics in ways that can be readily transported across state lines.

This is a time of great opportunity for mathematics education in the United States. Lines of communication have been opened among policy-makers, mathematicians, and mathematics educators, and changed educational policies provide the potential for educational improvement. Mathematicians have an essential role to play in fulfilling this potential in teacher education, curriculum, and assessment.

---

[40] *National Impact Report: Math and Science Partnership Program*, National Science Foundation, 2010, p. 15.

[41] In addition to the forthcoming CAEP standards, note the 2012 report *Supporting Implementation of the Common Core State Standards for Mathematics: Recommendations for Professional Development*, Friday Institute for Educational Innovation at the North Carolina State University College of Education.

[42] For an overview of MSP outcomes, including increases in student achievement, see *National Impact Report: Math and Science Partnership Program*, National Science Foundation, 2010, pp. 6, 10–12.

CHAPTER 3

Recommendations: Mathematics for Teachers; Roles for Mathematicians

This document's six recommendations are presented in two groups: the mathematics that teachers need to know; and mathematicians' roles in the mathematical education of teachers.

A. Mathematics for Teachers

The term "teacher of mathematics" includes early childhood and elementary-level generalist teachers as well as middle grades and high school teachers who teach mathematics classes. It also includes teachers of special needs students, English Language Learners, and other special groups, when those teachers have direct responsibility for teaching mathematics.

These recommendations are intended to apply to any pathway for teacher preparation and credentialing, regardless of form and type of institution, including undergraduate and graduate degree and certification programs, as well as any alternative program that prepares teachers of mathematics. It is assumed that this required coursework in mathematics is complemented with appropriate coursework in education, especially courses in methods of teaching mathematics. Implicit in these recommendations is that mathematicians and statisticians should teach, or co-teach, the mathematics discussed in these recommendations and that this instruction should occur at accredited institutions of higher education.[1]

Recommendation 1. Prospective teachers need mathematics courses that develop a solid understanding of the mathematics they will teach. The mathematical knowledge needed by teachers at all levels is substantial yet quite different from that required in other mathematical professions. Prospective teachers need to understand the fundamental principles that underlie school mathematics, so that they can teach it to diverse groups of students as a coherent, reasoned activity and communicate an appreciation of the elegance and power of the subject. Thus, coursework for prospective teachers should examine the mathematics they will teach in depth, from a teacher's perspective.

Recommendation 2. Coursework that allows time to engage in reasoning, explaining, and making sense of the mathematics that prospective teachers will teach is

---

[1]The recommendations for teacher preparation in this report are formulated in terms of courses and semester-hours, but this is not meant to exclude other ways of awarding credit or organizing teacher education. For example, collegiate institutions that do not follow a semester system with most courses earning 3 credit-hours will need to adapt these recommendations accordingly.

needed to produce well-started beginning teachers. Although the quality of mathematical preparation is more important than the quantity, the following recommendations are made for the amount of mathematics coursework for prospective teachers.

    i. *Prospective elementary teachers should be required to complete at least 12 semester-hours on fundamental ideas of elementary mathematics, their early childhood precursors, and middle school successors.*

    ii. *Prospective middle grades (5–8) teachers of mathematics should be required to complete at least 24 semester-hours of mathematics that includes at least 15 semester-hours on fundamental ideas of school mathematics appropriate for middle grades teachers.*

    iii. *Prospective high school teachers of mathematics should be required to complete the equivalent of an undergraduate major in mathematics that includes three courses with a primary focus on high school mathematics from an advanced viewpoint.*

Recommendations for the content and nature of this coursework are outlined in Chapters 4, 5, and 6 of this report.

At each level, these recommendations include courses especially designed for teachers. At the middle grades and high school levels, these recommendations also include courses such as calculus, linear algebra, and history of mathematics, which are designed for and taken by a wider undergraduate audience. The recommended statistics–probability courses are different from the statistics courses recommended in MET I because they focus on the data collection, analysis, and interpretation needed to teach the statistics outlined in the CCSS. Such courses are likely to be different from the more theoretically-oriented probability and statistics courses typically taken by science, technology, engineering, and mathematics (STEM) majors, and from the non-calculus-based statistics courses offered at many universities.

In states where teacher certification is accomplished as part of a post-baccalaureate program, a mathematics-intensive undergraduate major along with a minor in mathematics for teaching would be an acceptable preparation for the graduate degree in mathematics teaching. This graduate degree should include mathematics courses with a primary focus on high school mathematics from an advanced standpoint.

Although elementary certification in most states is still a K–6 and, in some states, a K–8 certification, state education departments and accreditation associations are urged to require all grades 5–8 teachers of mathematics to satisfy the 24-hour requirement recommended by this report.

Elementary mathematics specialists play important roles in elementary teachers' professional development. Those roles, and the education of specialists and early childhood teachers are discussed in Chapter 4.

Recommendation 3. Throughout their careers, teachers need opportunities for continued professional growth in their mathematical knowledge. Satisfying the minimum requirements for initial certification to teach mathematics does not ensure that even outstanding future teachers have the knowledge of mathematics, of teaching, and of students that is possessed by successful experienced teachers. Like all professionals, teachers need opportunities for professional growth throughout their careers. This need, however, takes on increased importance due to the wide adoption of the CCSS.

In-service programs should offer teachers content-based professional growth at levels appropriate for their experience, as they make the transition from new teacher, to mid-career professional, to master teacher. Opportunities for mathematical growth should include school- and district-based professional development, university-based graduate courses and "short courses" (e.g., one- or two-week intensive courses), teacher-driven professional experiences (e.g., lesson study), and teacher–mathematician partnerships (e.g., math teachers' circles).[2] There is an important role for mathematicians in all these activities.

Regular opportunities to learn mathematics beyond preparation courses are particularly important at the high school level. A reasonable goal for initial certification at this level is to create beginning teachers who are able to teach competently a portion of the high school curriculum and who are prepared to learn throughout their careers from their teaching and professional development experiences. Most well-prepared new high school teachers will be ready to teach algebra and geometry. But, most new high school teachers will require further coursework to be well prepared to teach subjects such as precalculus, calculus, discrete mathematics, matrix algebra, and more than basic statistics.

Recommendation 4. All courses and professional development experiences for mathematics teachers should develop the habits of mind of a mathematical thinker and problem-solver, such as reasoning and explaining, modeling, seeing structure, and generalizing. Courses should also use the flexible, interactive styles of teaching that will enable teachers to develop these habits of mind in their students. This recommendation is at least as important for practicing teachers as future teachers. A worthy goal of mathematics instruction for any undergraduate is to develop not only knowledge of content but also the ability to work in ways characteristic of the discipline. For teachers, this is not only worthy, but necessary. In order to develop these abilities in their students, teachers must experience them in their own mathematical education, through, for example, immersion experiences, research projects, or seminars devoted to doing mathematics. To help their students achieve the CCSS Standards for Mathematical Practice, teachers must not only understand the practices of the discipline, but how these practices can occur in school mathematics and be acquired by students.

B. Roles for Mathematicians in Teacher Education

Recommendation 5. At institutions that prepare teachers or offer professional development, teacher education must be recognized as an important part of a mathematics department's mission and should be undertaken in collaboration with mathematics education faculty. More mathematics faculty need to become deeply involved in PreK–12 mathematics education by participating in preparation and professional development for teachers and becoming involved with local schools or districts. Mathematics departments need to encourage and reward faculty for these efforts. Departments also need to devote commensurate resources to designing and staffing courses

---

[2]Lesson study is a process in which teachers jointly plan, observe, analyze, and refine actual classroom lessons. Math teachers' circles focus primarily on giving teachers an experience to be learners and doers of mathematics. See the web resources for further information and examples.

for prospective and practicing teachers. For some departments, these courses may require new institutional arrangements. Within a department, courses designed for prospective high school teachers can serve the needs of other mathematics majors to work in ways characteristic of mathematics. Courses designed for practicing teachers can combine intensive multi-day meetings with distance learning. One course can be shared by several institutions, or as part of a regional mathematics education consortium.[3] (See the web resources for examples.)

At a minimum, oversight of programs for teachers should be the responsibility of a faculty member with expertise in teacher education as well as mathematics. As for any program, continuity is desirable, for both administrators and instructors. For administrators, this may be of special importance when it requires coordination among academic units (e.g., mathematics, statistics, and education) or with school personnel. For instructors, continuity—and support for professional development— afford increased expertise in teaching teachers.

Although most statistics courses for future teachers are taught by mathematicians or statisticians in mathematics departments, on campuses where there is a separate department of statistics, statistics courses for teachers are seldom a department priority.[4] This needs to change. Statements for mathematics departments also apply to statistics departments who are responsible for the statistical education of future mathematics teachers.

State departments of education and local school districts recognize the need for continuing education and implement policies requiring professional development or graduate education.[5] Unfortunately, little of either has been content-based. At the same time, few mathematics or statistics departments provide any opportunities at the graduate level designed to meet the professional needs of PreK–12 mathematics teachers. More mathematics and statistics departments need to have faculty members who work with educators, teachers, and school district personnel to design and implement content-based professional development in schools, districts, and states. National and regional efforts are needed to help prepare these faculty members to contribute effectively to professional development activities for teachers.

There are notable exceptions that can serve as models for departments interested in supporting and serving this important part of the mathematics education community. Examples can be found in the web resources for this report.

Recommendation 6. Mathematicians should recognize the need for improving mathematics teaching at all levels. Mathematics education, including the mathematical education of teachers, can be greatly strengthened by the growth of a mathematics education community that includes mathematicians as one of many constituencies committed to working together to improve mathematics instruction at all levels and to raise professional standards in teaching. It is important to encourage partnerships between mathematics faculty and mathematics education faculty, between

---

[3]See Recommendation 13 of *National Task Force on Teacher Education in Physics: Report Synopsis*, American Association of Physics Teachers, the American Physical Society, & the American Institute of Physics, 2010.

[4]In the 2005 CBMS survey, special courses for K–8 teachers were offered by 11% of Ph.D.-granting and 33% of M.A.-granting statistics departments. Less than 0.5% of statistics departments surveyed reported that special sections of regular courses were designated for K–8 teachers. See Table SP.3.

[5]For an overview, see *Key State Education Policies on PK–12 Education: 2008*, Council of Chief State School Officers, p. 22.

faculty in two- and four-year institutions, and between mathematics faculty and school mathematics teachers, as well as state, regional, and school-district leaders.

In particular, as part of the mathematics education community, mathematicians should support the professionalism of mathematics teachers by:

  i. endeavoring to ensure that PreK 12 mathematics teachers have sufficient knowledge and skills upon receiving initial certification;

 ii. encouraging all who teach mathematics to strive for continual improvement in their mathematics teaching;

iii. joining with teachers at different levels to learn with and from each other.

Strategies for raising professional standards in teaching include developing a career ladder that keeps outstanding teachers in the profession, and providing professional development opportunities for teachers to grow from early career teachers, to mid-career teachers, to master teachers. Although mathematics teachers themselves need to provide leadership, this effort will be enhanced by the development of a comprehensive professional community involving all who teach mathematics or statistics. Society will be better served by focusing efforts on institutional arrangements and professional practices that foster expertise, such as on raising standards at the time of entry into school teaching and providing professional development based on content, rather than relying on punitive approaches focused on weeding out ineffective practicing teachers.

There are many initiatives, communities, and professional organizations focused on aspects of building professionalism in mathematics teaching. More explicit efforts are needed to bridge current communities in ways that build upon mutual respect and the recognition that these initiatives provide opportunities for professional growth for higher education faculty in mathematics, statistics, and education as well as for the mathematics teachers, coaches, and supervisors in the PreK–12 community. The web resources for this report include examples of such collaborative work. Also needed are more opportunities for observation and discussion of the work of teaching, including professional learning communities, math teachers' circles, conferences, and publications, from newsletters to scholarly articles. Mathematicians have an important role to play in all these efforts.

Finally, becoming part of a community that connects all levels of mathematics education will offer mathematicians more opportunities to participate in setting standards for accreditation of teacher preparation programs and for teacher certification via standard and alternative pathways.

CHAPTER 4

Elementary Teachers

What mathematics should future elementary teachers study to prepare for their careers? What mathematics coursework and programs will prepare elementary teachers for teaching mathematics? What sorts of professional development experiences will develop and sustain high quality mathematics teaching in elementary school? How can mathematicians make valuable contributions to these endeavors? These questions are the topics of this chapter. Coursework in mathematical pedagogy is assumed to be part of a preparation program, but is not discussed in detail.

In this chapter, the term "elementary teacher" is defined as a teacher who teaches mathematics at the K–5 level.[1]

A major advance in teacher education is the realization that teachers should study the mathematics they teach in depth, and from the perspective of a teacher. There is widespread agreement among mathematics education researchers and mathematicians that it is not enough for teachers to rely on their past experiences as learners of mathematics.[2] It is also not enough for teachers just to study mathematics that is more advanced than the mathematics they will teach. Importantly, mathematics courses and professional development for elementary teachers should not only aim to remedy weaknesses in mathematical knowledge, but also help teachers develop a deeper and more comprehensive view and understanding of the mathematics they will or already do teach.

Thus, this report recommends that before beginning to teach, an elementary teacher should study in depth, and from a teacher's perspective, the vast majority of K–5 mathematics, its connections to prekindergarten mathematics, and its connections to grades 6–8 mathematics. By itself, this expectation is not sufficient to guarantee high quality teaching. In particular, teachers will also need courses in

---

Note that the MET II web resources at www.cbmsweb.org give URLs for the CCSS, the Progressions for the CCSS, and other relevant information.

[1]As noted in Chapter 3, "Although elementary certification in most states is still a K–6 and, in some states, a K–8 certification, state education departments and accreditation associations are urged to require all grades 5–8 teachers of mathematics to satisfy the 24-hour requirement recommended by this report." Chapters 4 and 5 allow for a period of transition.

[2]For example, "It is widely assumed—some would claim common sense—that teachers must know the mathematical content they teach" (*Foundations for Success: Reports of Task Groups of the National Mathematics Advisory Panel*, 2008, p. 5-6). "Aspiring elementary teachers must begin to acquire a deep conceptual knowledge of the mathematics that they will one day need to teach, moving well beyond mere procedural understanding" (*No Common Denominator*, 2008, National Council on Teacher Quality). "Mathematics courses for future teachers should develop 'deep understanding' of mathematics, particularly of the mathematics taught in schools at their chosen grade level" (Curriculum Foundations Project, 2001, Mathematical Association of America). See also *Preparing Teachers: Building Sound Evidence for Sound Policy*, 2010, National Research Council, p. 123.

mathematical pedagogy. However, there is no substitute: a strong understanding of the mathematics a teacher will teach is necessary for good teaching. Every elementary student deserves a teacher who knows, very well, the mathematics that the student is to learn. As reasonable as this expectation may seem, it is not routinely achieved.[3]

With the advent of the Common Core State Standards for Mathematics (CCSS), there is now a succinct description of the mathematics to be taught and learned at the elementary school level in the United States. The CCSS describe not only the specific mathematical skills and understandings that students are to acquire but also the kinds of mathematical practice that students are to develop.

Several points about the CCSS Standards for Mathematical Practice bear emphasizing. First, although those standards were written for K–12 students, they apply to all who do mathematics, including elementary teachers. Second, the features of mathematical practice described in these standards are not intended as separate from mathematical content. Teachers should acquire the types of mathematical expertise described in these standards as they learn mathematics. And finally, *engaging in mathematical practice takes time and opportunity*, so that coursework and professional development for teachers must be planned with that in mind. Time and opportunity to think about, discuss, and explain mathematical ideas are essential for learning to treat mathematics as a sense-making enterprise.

Readers who are new to the preparation and professional development of elementary teachers may find some of the ideas, examples, and terms (e.g., "unit rate," "tape diagram") presented in this chapter unfamiliar or unusual. Interested readers, and those who will teach mathematics courses and provide professional development for teachers, should consult additional sources for definitions and examples, including the CCSS and the Progressions for the CCSS (see the web resources associated with this report). Materials that have been carefully designed for courses and professional development opportunities for teachers exist and are a sensible starting point for those who will begin teaching such courses and providing professional development.

What kinds of problems might prospective or practicing elementary teachers work on in coursework or in professional development experiences? What kinds of mathematical discussions, explanations, and thinking might they engage in? The first section of this chapter gives very brief sketches of how the mathematics might be treated in coursework or professional development for teachers, showing its difference from the content of courses often taken by teachers, e.g., college algebra.

The second section of this chapter suggests how this mathematics can be organized in courses, programs, or seminars for prospective or practicing elementary teachers. In addition, this section describes other types of professional development for teachers that afford opportunities for mathematicians to participate in the broader mathematics education community. The final sections of the chapter

---

[3]An international comparison of prospective elementary teachers found that 48% of the U.S. teachers did not score above "Anchor point 2." Teachers with this score often had trouble reasoning about factors, multiples, and percentages. See Tatto & Senk, "The Mathematics Education of Future Primary and Secondary Teachers: Methods and Findings from the Teacher Education and Development Study in Mathematics," *Journal of Mathematics Teacher Education*, 2011, pp. 129–130. *Preparing Teachers* discusses concern about the adequacy of current teacher preparation in mathematics, especially for elementary teachers. See Chapter 6, especially p. 124.

discuss the preparation and professional development of elementary mathematics specialists, early childhood teachers, and teachers of special populations.

Essential Grades K–5 Ideas for Teachers

This section uses the CCSS as a framework for outlining the mathematical ideas that elementary teachers, both prospective and practicing, should study and know. The CCSS standards for mathematical content are organized into clusters of related standards and the clusters are organized into mathematical domains, which span multiple grade levels (see Appendix B). Brief descriptions of how the mathematics of each domain progresses across grade levels and is connected within or across grades to standards in other domains appear in the Progressions for the CCSS (see the web resources associated with this report).

Because elementary teachers prepare their students for the middle grades, courses and seminars for elementary teachers should also attend to how the mathematical ideas of the elementary grades build to those at the middle grades, and should highlight connections between topics at the elementary and middle levels. Thus, courses and professional development will need to devote time to ideas within the middle grades domains of Ratio and Proportional Relationships, The Number System, Expressions and Equations, and Statistics and Probability (see Chapter 5).

This section lists essential ideas of each K–5 domain and important connections to prior or later grades that teachers need to know well. These listings are not intended as comprehensive; and instructors are encouraged to refer to the CCSS, related progressions, and other references given in the web resources for further details and discussion.

For each domain, the list of essential ideas is followed by a list of related activities that could be used in teacher preparation or professional development.

A given activity may provide opportunities to demonstrate or develop various kinds of expertise described by one or more of the CCSS standards for mathematical practice. These are indicated by the number and heading of the associated standard. For example, "MP 1 Make sense of problems and persevere in solving them" indicates expertise connected with the first Standard for Mathematical Practice might be used. (The full text for all eight Standards for Mathematical Practice appears as Appendix C of this report.) Note that although a particular activity might provide opportunities to use or increase expertise, instructors should expect to foster engagement in these opportunities. Also, instructors might periodically remind teachers to review and reflect on the Standards for Mathematical Practice so that they become more familiar with the types of expertise described by these standards in the context of elementary mathematics.

Counting and Cardinality (Kindergarten)

- The intricacy of learning to count, including the distinction between counting as a list of numbers in order and counting to determine a number of objects.

Illustrative activity:

> Examine counting errors that young children typically make and study the
> learning path of counting.[4] (This includes connections to prekindergarten
> mathematics.)

> MP 2 Reason abstractly and quantitatively.
> MP 4 Model with mathematics.

Operations and Algebraic Thinking (Kindergarten–Grade 5)
- The different types of problems solved by addition, subtraction, multiplication, and division, and meanings of the operations illustrated by these
  problem types.[5]
- Teaching–learning paths for single-digit addition and associated subtraction and single-digit multiplication and associated division, including the
  use of properties of operations (i.e., the field axioms).
- Recognizing the foundations of algebra in elementary mathematics, including understanding the equal sign as meaning "the same amount as"
  rather than a "calculate the answer" symbol.

Illustrative activities:

(1) Write equations for addition and subtraction problems of different types
    and determine which cases have a "situation equation" (an equation that
    fits naturally with the wording of the problem) that is different from a
    "solution equation" (an equation that is especially helpful for solving the
    problem).

    MP 2 Reason abstractly and quantitatively.

(2) Recognize that commutativity for multiplication is not obvious and use
    arrays to explain why multiplication is commutative.

    MP 3 Construct viable arguments and critique the reasoning of others.
    MP 5 Use appropriate tools strategically.

(3) Explain why we can't divide by 0.

    MP 3 Construct viable arguments and critique the reasoning of others.
    MP 7 Look for and make use of structure.

(4) Explore and discuss the different ways remainders can be interpreted when
    solving division problems.

    MP 4 Model with mathematics.
    MP 6 Attend to precision.

(5) Explain how to solve equations such as $283 + 19 = x + 18$ by "thinking
    relationally" (e.g., by recognizing that because 19 is 1 more than 18, x
    should be 1 more than 283 to make both sides equal) rather than by
    applying standard algebraic methods.

---

[4]See the National Research Council report *Mathematics Learning in Early Childhood: Paths
Toward Excellence and Equity* and the Counting and Cardinality Progression.

[5]See CCSS, pp. 88–89; or the Operations and Algebraic Thinking Progression for details and
examples of situation and solution equations.

MP 3 Construct viable arguments and critique the reasoning of others.
MP 7 Look for and make use of structure.
MP 8 Look for and express regularity in repeated reasoning.

Number and Operations in Base Ten (Kindergarten–Grade 5)

- How the base-ten place value system relies on repeated bundling in groups of ten and how to use objects, drawings, layered place value cards, and numerical expressions to help reveal base-ten structure. Developing progressively sophisticated understandings[6] of base-ten structure as indicated by these expressions:

$$357 = 300 + 50 + 7$$
$$= 3 \times 100 + 5 \times 10 + 7 \times 1$$
$$= 3 \times (10 \times 10) + 5 \times 10 + 7 \times 1$$
$$= 3 \times 10^2 + 5 \times 10^1 + 7 \times 10^0.$$

- How efficient base-ten computation methods for addition, subtraction, multiplication, and division rely on decomposing numbers represented in base ten according to the base-ten units represented by their digits and applying (often informally) properties of operations, including the commutative and associative properties of addition and the distributive property, to decompose the calculation into parts. How to use math drawings or manipulative materials to reveal, discuss, and explain the rationale behind computation methods.
- Extending the base-ten system to decimals and viewing decimals as address systems on number lines. Explaining the rationales for decimal computation methods. (This includes connections to grades 6–8 mathematics.)

Illustrative activities:

(1) Make simple base-ten drawings to calculate $342 - 178$ and identify correspondences with numerical written methods. Compare advantages and disadvantages to each numerical written variation.

   MP 1 Make sense of problems and persevere in solving them.
   MP 2 Reason abstractly and quantitatively.

(2) Examine hypothetical or actual student calculation methods and decide if the methods are valid or not. For example, recognize that if a student calculates 23×45 by calculating 20×40 and 3×5 and adding the two results, the method is not legitimate but can be modified to become correct by adding the two missing products that arise from applying the distributive property, which can also be seen in an area or array model.

   MP 3 Construct viable arguments and critique the reasoning of others.
   MP 7 Look for and make use of structure.

---

[6]For examples of how teachers may construe the base-ten system, see Thanheiser, "Pre-service Elementary School Teachers' Conceptions of Multidigit Whole Numbers," *Journal for Research in Mathematics Education*, 2009.

(3) Explain how to use properties of operations to make some calculations such as 98×15 or 24×25 easy to carry out mentally and write strings of equations, such as $24 \times 25 = (6 \times 4) \times 25 = 6 \times (4 \times 25) = 6 \times 100 = 600$, to show how properties of operations support the "mental math."

MP 3 Construct viable arguments and critique the reasoning of others.
MP 7 Look for and make use of structure.

Number and Operations—Fractions (Grades 3–5)

- Understanding fractions as numbers which can be represented with lengths and on number lines. Using the CCSS development of fractions to define fractions a/b as a parts, each of size $1/b$. Attending closely to the whole (referent unit) while solving problems and explaining solutions.
- Recognizing that addition, subtraction, multiplication, and division problem types and associated meanings for the operations (e.g., CCSS, pp. 88–89) extend from whole numbers to fractions.
- Explaining the rationale behind equivalent fractions and procedures for adding, subtracting, multiplying, and dividing fractions. (This includes connections to grades 6–8 mathematics.)
- Understanding the connection between fractions and division, $a/b = a \div b$, and how fractions, ratios, and rates are connected via unit rates. (This includes connections to grades 6–8 mathematics. See the Ratio and Proportion Progression for a discussion of unit rate.)

Illustrative activities:

(1) Use drawings and reasoning to solve problems and explain solutions. For example:

> One serving of rice is $1/2$ cup. You ate $2/3$ of a cup of rice. How many servings did you eat?

Examine and critique reasoning, such as:

> A student said that $2/3$ of a cup of rice is 1 serving plus another $1/6$. Is that correct? [It is 1 serving plus another $1/6$ of a cup of rice, but the $1/6$ of a cup of rice is $1/3$ of a serving. That is because $1/2 = 3/6$. The $1/6$ of a cup of rice is one of the 3 sixths of a cup that make a $1/2$ cup serving.]

MP 1 Make sense of problems and persevere in solving them.
MP 2 Reason abstractly and quantitatively.
MP 3 Construct viable arguments and critique the reasoning of others.
MP 5 Use appropriate tools strategically.

(2) Give rationales underlying methods for comparing fractions, including comparing fractions with common denominators or common numerators and explain how to compare fractions by relating them to benchmarks such as $1/2$ or 1. For example, use reasoning to compare $73/74$ and $85/86$.

MP 2 Reason abstractly and quantitatively.
MP 3 Construct viable arguments and critique the reasoning of others.
MP 7 Look for and make use of structure.

(3) Explain how it can happen that the multiplication of fractions can produce a product that is smaller that its factors and division of fractions can produce a quotient that is larger than divisor and dividend.

MP 2 Reason abstractly and quantitatively.

(4) Calculate percentages mentally and write equations to show the algebra behind the mental methods, such as calculating 45% of 120 by taking half of 120, which is 60, then taking away 10% of that, leaving 54.

$$\begin{aligned}
\text{Possible equations: } 45\% \cdot 120 &= (50\% - 5\%)120 \\
&= 50\% \cdot 120 - 5\% \cdot 120 \\
&= 60 - (10\% \cdot \tfrac{1}{2})120 \\
&= 60 - 10\%(\tfrac{1}{2} \cdot 120) \\
&= 60 - 10\% \cdot 60 \\
&= 60 - 6 = 54.
\end{aligned}$$

MP 2 Reason abstractly and quantitatively.
MP 3 Construct viable arguments and critique the reasoning of others.
MP 7 Look for and make use of structure.

Measurement and Data (Kindergarten–Grade 5)
- The general principles of measurement, the process of iterations, and the central role of units: that measurement requires a choice of measureable attribute, that measurement is comparison with a unit and how the size of a unit affects measurements, and the iteration, additivity, and invariance used in determining measurements.
- How the number line connects measurement with number through length (see the Geometric Measurement Progression).
- Understanding what area and volume are and giving rationales for area and volume formulas that can be obtained by finitely many compositions and decompositions of unit squares or unit cubes, including formulas for the areas of rectangles, triangles, and parallelograms, and volumes of rectangular prisms. (This includes connections to grades 6–8 geometry, see the Geometric Measurement Progression.)
- Using data displays to ask and answer questions about data. Understanding measures used to summarize data, including the mean, median, interquartile range, and mean absolute deviation, and using these measures to compare data sets. (This includes connections to grades 6–8 statistics, see the Measurement Data Progression.)

Illustrative activities:
(1) Explore the distinction and relationship between perimeter and area, such as by fixing a perimeter and finding the range of areas possible or by fixing an area and finding the range of perimeters possible.

MP 2 Reason abstractly and quantitatively.
MP 3 Construct viable arguments and critique the reasoning of others.
MP 8 Look for and express regularity in repeated reasoning.

(2) Investigate whether the area of a parallelogram is determined by the lengths of its sides. Given side lengths, which parallelogram has the largest area? Explain how to derive the formula for the area of a parallelogram, including for "very oblique" cases, by decomposing and recomposing parallelograms and relating their areas to those of rectangles.

MP 3 Construct viable arguments and critique the reasoning of others.
MP 7 Look for and make use of structure.
MP 8 Look for and express regularity in repeated reasoning.

(3) Examine the distinction between categorical and numerical data and reason about data displays. For example:

> Given a bar graph displaying categorical data, could we use the mean of the frequencies of the categories to summarize the data? [No, this is not likely to be useful.]

> Given a dot plot displaying numerical data, can we calculate the mean by adding the frequencies and dividing by the number of dots? [No, this is like the previous error.]

MP 3 Construct viable arguments and critique the reasoning of others.
MP 5 Use appropriate tools strategically.

(4) Use reasoning about proportional relationships to argue informally from a sample to a population. For example:

> If 10 tiles were chosen randomly from a bin of 200 tiles (e.g., by selecting the tiles while blindfolded), and if 7 of the tiles were yellow, then about how many yellow tiles should there be in the bin? Imagine repeatedly taking out 10 tiles until a total of 200 tiles is reached. What does this experiment suggest? Then investigate the behavior of sample proportion by taking random samples of 10 from a bin of 200 tiles, 140 of which are yellow (replacing the 10 tiles each time). Plot the fraction of yellow tiles on a dot plot or line plot and discuss the plot. How might the plot be different if the sample size was 5? 20? Try these different sample sizes.

MP 2 Reason abstractly and quantitatively.
MP 4 Model with mathematics.
MP 5 Use appropriate tools strategically.

Geometry (Kindergarten–Grade 5)

- Understanding geometric concepts of angle, parallel, and perpendicular, and using them in describing and defining shapes; describing and reasoning about spatial locations (including the coordinate plane).
- Classifying shapes into categories and reasoning to explain relationships among the categories.
- Reason about proportional relationships in scaling shapes up and down. (This is a connection to grades 6–8 geometry.)

Illustrative activities:

(1) Explore how collections of attributes are related to categories of shapes. Sometimes, removing one attribute from a collection of attributes does not change the set of shapes the attributes apply to and sometimes it does.

MP 7 Look for and make use of structure.

(2) Reason about scaling in several ways: If an 18-inch by 72-inch rectangular banner is scaled down so that the 18-inch side becomes 6 inches, then what should the length of the adjacent sides become? Explain how to reason by:

Comparing the 18-inch and 6-inch sides. [The 18-inch side is 3 times the length of the 6-inch side, so the same relationship applies with the 72-inch side and the unknown side length.]

Comparing the 18-inch and 72-inch sides. [The 72-inch side is 4 times the length of the 18-inch side, so the unknown side length is also 4 times the length of the 6-inch side.]

MP 2 Reason abstractly and quantitatively.
MP 4 Model with mathematics.
MP 7 Look for and make use of structure.

The Preparation and Professional Development of Elementary Teachers

The mathematics of elementary school is full of deep and interesting ideas, which can be studied repeatedly, with increasing depth and attention to detail and nuance. Therefore, although prospective teachers will undertake an initial study of elementary mathematics from a teacher's perspective in their preparation program, practicing teachers will benefit from delving more deeply into the very same topics. Perhaps surprisingly, mathematics courses that explore elementary school mathematics in depth can be genuinely college-level intellectual experiences, which can be interesting for instructors to teach and for teachers to take.[7]

PROGRAMS FOR PROSPECTIVE TEACHERS

Programs designed to prepare elementary teachers should include 12 semester-hours focused on a careful study of mathematics associated with the CCSS (K–5 and related aspects of 6–8 domains) from a teacher's perspective. This includes, but is not limited to studying all the essential ideas described in the previous section and their connections with the essential ideas of grades 6–8 described in Chapter 5. It also includes some attention to methods of instruction. Number and operations, treated algebraically with attention to properties of operations, should occupy about 6 of those hours, with the remaining 6 hours devoted to additional ideas of algebra (e.g., expressions, equations, sequences, proportional relationships, and linear relationships), and to measurement and data, and to geometry.

---

[7]Beckmann, "The Community of Math Teachers, from Elementary School to Graduate School," *Notices of the American Mathematical Society*, 2011.

When possible, program designers should consider courses that blend the study of content and methods. Prospective teachers who have a limited mathematical background will need additional coursework in mathematics.

It bears emphasizing that familiar mathematics courses such as college algebra, mathematical modeling, liberal arts mathematics, and even calculus or higher level courses *are not an appropriate substitute for the study of mathematics for elementary teachers*, although they might make reasonable additions.[8] Also, it is unlikely that knowledge of elementary mathematics needed for teaching can be acquired through experience in other professions, even mathematically demanding ones.

PROFESSIONAL DEVELOPMENT FOR PRACTICING TEACHERS

Once they begin teaching, elementary teachers need continuing opportunities to deepen and strengthen their mathematical knowledge for teaching, particularly as they engage with students and develop better understanding of their thinking.

Professional development may take a variety of forms. A group of teachers might work together in a professional learning community, and they might choose to focus deeply on one topic for a period of time. For example, the teachers at the same grade level in several schools might spend a term studying fractions in the CCSS, the grade 3–5 Fractions Progression and other curriculum documents, followed by designing, teaching, and analyzing lessons on fraction multiplication using a lesson study format.[9] Or a group of teachers who teach several grade levels at one school might meet regularly to study how related topics progress across grade levels. A group of teachers might watch demonstration lessons taught by a mathematics specialist and then meet to discuss the lessons, plan additional lessons, and study the mathematics of the lessons.[10] Teachers might also complete mathematics courses specifically designed as part of a graduate program for elementary teachers. Professional development can take place at school, either during or after school hours, or on college campuses after school hours or during the summer. However it is organized, as discussed in Chapter 2, the best professional development is ongoing, directly relevant to the work of teaching mathematics, and focused on mathematical ideas.

Regardless of format, as part of a professional development program, teachers could study mathematics materials specifically designed for professional development and, if the textbook series is carefully designed, the teacher's guides for the mathematics textbooks used at their schools. Mathematics specialists or college-based mathematics educators or mathematicians might lead sessions in which they engage teachers in solving problems, thinking together, and discussing mathematical ideas. Teachers could bring student work to share and discuss with the group. Opportunities to examine how students are thinking about mathematical ideas,

---

[8]For instance, a study of prospective elementary and secondary teachers found that many either did not know that division by 0 was undefined or were unable to explain why it was undefined. On average, the secondary teachers had taken over 9 college-level mathematics courses. Ball, "Prospective Elementary and Secondary Teachers' Understanding of Division," *Journal for Research in Mathematics Education*, 1990.

[9]Lesson study is a process in which teachers jointly plan, observe, analyze, and refine actual classroom lessons.

[10]See this chapter's section on mathematics specialists for more discussion about their roles in professional development for teachers.

and to learn about learning paths and tasks designed to help students progress along learning paths are especially important for elementary teachers and can lead to improved student outcomes. Together, teachers could write problems for their students that they design to get a sense of what students already know about an upcoming topic of instruction (an example of formative assessment). They could share results of assessments and, based on the outcome, plan appropriate tasks for the students. Throughout, outside experts, such as mathematicians, statisticians and mathematics educators in higher education or professionals from mathematically-intensive fields could work with the teachers to bring a fresh perspective and to help teachers go deeply into the content. A side benefit of this work to those in higher education is the opportunity to think about undergraduate mathematics teaching and the connection between college-level mathematics courses and K–12 education.

Engaging in mathematical practice. Teacher preparation and professional development can provide opportunities to *do* mathematics and to develop mathematical habits of mind. Teachers must have time, opportunity, and a nurturing environment that encourages them to make sense of problems and persevere in solving them. They should experience the enjoyment and satisfaction of working hard at solving a problem so that they realize this sort of intellectual work can be satisfying and so that they don't seek to shield their students from the struggles of learning mathematics. Teachers should have time and opportunity to reason abstractly and quantitatively, to construct viable arguments, to listen carefully to other people's reasoning, and to discuss and critique it. Some teachers may not realize that procedures and formulas of mathematics can be explained in terms of more fundamental ideas and that deductive reasoning is considered an essential part of mathematics. Teachers should have the opportunity to model with mathematics and to *mathematize* situations by focusing on the mathematical aspects of a situation and formulating them in mathematical terms. Elementary teachers should know ways to use mathematical drawings, diagrams, manipulative materials, and other tools to illuminate, discuss, and explain mathematical ideas and procedures. Teachers need practice being precise and deliberate when they discuss their reasoning, and to be on the lookout for incomplete or invalid arguments. Especially important is that teachers learn to use mathematical terminology and notation correctly. And finally, teachers need opportunities to look for and use regularity and structure by seeking to explain the phenomena they observe as they examine different solution paths for the same problem.

Use technology and other tools strategically. Since the publication of MET I, the technology available to support the teaching and learning of mathematics has changed dramatically. These tools include interactive whiteboards and tablets, mathematics-specific technology such as virtual manipulatives and "quilting" software, and an ever-expanding set of applets, apps, web sites, and multimedia materials. Thus, it is important that teacher preparation and professional development programs provide opportunities for teachers to use these tools in their own learning so that they simultaneously advance their mathematical thinking, expand the repertoire of technological tools with which they are proficient, and develop an awareness of the limitations of technology. Teachers should have experiences using technology as a computational and problem solving tool. When technology is used as a computational tool, learners use it to perform a calculation or produce a graph or table in order to use the result as input to analyze a mathematical situation. They should also learn to use technology as a problem solving tool, or to conduct an investigation by taking a deliberate mathematical action, observing the consequences, and reflecting on the mathematical implications of the consequences. Teachers must have opportunities to engage in the use of a variety of technological tools, including those designed for mathematics and for teaching mathematics, to explore and deepen their understanding of mathematics, even if these tools are not the same ones they will eventually use with children.

Technology is one of many tools available for learning and teaching mathematics. Others are traditional tools of teaching such as blackboards.[11] Some are manipulative materials such as base-ten blocks, which can be used for early work with place value and operations with whole numbers and decimals; pattern blocks, which can be used for work with fractions; tiles; and counters. Teachers need to develop the ability to critically evaluate the affordances and limitations of a given tool, both for their own learning and to support the learning of their students. In mathematics courses for teachers, instructors should model successful ways of using such tools, and provide opportunities to discuss mathematical issues that arise in their use.

Challenges in the Education of Elementary Teachers

Prospective elementary school teachers frequently come to their teacher preparation programs with their own views about what it means to know and do mathematics and how it is learned. They sometimes feel insecure about their own mathematical knowledge while believing that learning to teach is a matter of learning to explain procedures clearly and assembling a toolkit of tasks and activities to use with children. As discussed in Chapter 2, some teachers may have a "fixed mind-set" about learning rather than a "growth mind-set" and may not recognize that everyone can improve their capacity to learn and understand mathematics. Instructors need to recognize that the messages of their courses and professional development opportunities may be filtered through such views. Some prospective teachers, although they may not like mathematics or feel confident in their ability to do it, do not think they need to learn more mathematics. In particular, they do not think there is anything else for them to learn about the content of elementary school mathematics. Similarly, prospective and practicing teachers may not be familiar with all of the content and practices outlined in the CCSS. Thus, they may question the need to learn these things in their teacher preparation programs and professional development and may actively resist and reject such instruction. Instructors may need to spend time focusing on the importance of not only a productive disposition toward mathematics,[12] but a recognition of the depth and importance of elementary mathematics, explaining the rationale for its structure and content, and its relationship with the preparation or professional development program.

Responsibility for designing and running elementary teacher preparation programs generally resides in colleges of education, and with faculty members whose

---

[11]See, e.g., discussion of the use and organization of the blackboard in Lewis, *Lesson Study*, Research for Better Schools, 2002, pp. 97–98.

[12]"Productive disposition" is discussed in the National Research Council report *Adding It Up*, pp. 116–117, 131–133.

primary focus and expertise is not mathematics. These faculty members face increasing pressure to add courses related to English Language Learners, special education, educational policy, assessment, and other contemporary issues, which sometimes leads to the elimination or reduction of mathematics courses for prospective
teachers. Faculty may also get push-back from pre-service teachers who do not see
the value of the mathematics courses they are required to take. Thus, it is important
for those who are concerned with the mathematical preparation of teachers to be in
close contact with the faculty who make decisions about the preparation program
to educate them about the need for strong mathematical preparation for elementary teachers. Reciprocally, those advocating for the mathematical preparation of
teachers need to be well-informed about contemporary issues such as those noted
above and thoughtful about how these issues might be addressed in mathematics
and mathematics education courses.

Few people trained as mathematicians have thought deeply about how courses
for prospective or practicing elementary school teachers might be taught, and
there is little support, professional development, or on-the-job training available
for them.[13] In some cases, mathematicians do not see the deep study of elementary
mathematics content as worthy of college credit. They may try to make the course
content "harder" by introducing higher-level mathematics or teach it as a skills
course. Or they may ask elementary teachers to take courses such as calculus or
other college mathematics courses in lieu of courses on elementary mathematics.
In contrast, the content outlined in the previous section shows that there is much
to be taught and learned. Colleges and universities need to provide support for
those teaching this content to develop their understanding of the manner in which
it should be taught.

Practicing teachers may feel overwhelmed by the burdens, mandates, and ac-
countability structures imposed on them by their schools, districts, and states.
Teachers in professional development seminars may need some time to communicate with each other about these problems before they turn to more specific thinking
about mathematics and its instruction. Professional developers must be sensitive
to the pressures that teachers face while also making productive use of valuable
time for teachers to think about mathematics and its teaching.[14]

Elementary Mathematics Specialists

Increasingly, school districts have utilized mathematics specialists at the elementary school level.[15] Within their schools, mathematics specialists are regarded
as experts. Administrators and other teachers rely upon them for guidance in
curriculum selection, instructional decisions, data analysis, teacher mentoring in
mathematics, communication with parents, and a host of other matters related
to the teaching and learning of mathematics. Depending on location, a specialist

---

[13]See discussion of support in Masingila et al., "Who Teaches Mathematics Content Courses
for Prospective Elementary Teachers in the United States? Results of a National Survey," *Journal
of Mathematics Teacher Education*, 2012.

[14]See, e.g., Schoenfeld, "Working with Schools: The Story of a Mathematics Education
Collaboration," *American Mathematical Monthly*, 2009, p. 202.

[15]See Fennell, "We Need Elementary Mathematics Specialists Now, More Than Ever: A Historical Perspective and Call to Action," *National Council of Supervisors of Mathematics Journal*,
2011.

may hold the title elementary mathematics coach, elementary mathematics instructional leader, mathematics support teacher, mathematics resource teacher, mentor teacher, or lead teacher.[16] Specialists serve a variety of functions: mentoring their teacher colleagues, conducting professional development, teaching demonstration lessons, leading co-planning or data teams sessions, observing teachers, or serving as the lead teacher for all of the mathematics classes for a particular group of students.

In several states, specialists and mathematicians collaborate in teaching courses offered for teachers in the specialist's district. Because the specialists remain in their districts, they are able to sustain teachers' learning after the courses. This strategy has been successful in improving student learning.[17]

In 2009, the Association of Mathematics Teacher Educators developed standards for elementary mathematics specialists (EMS), drawing on MET I and other reports.[18] In addition to an understanding of the content in grades K–8, these standards call for EMS to be prepared in the areas of learners and learning (including teachers as adult learners), teaching, and curriculum and assessment. Further, EMS are asked to develop knowledge and skills in the area of leadership as they are often called upon to function in a leadership capacity at the building or district level.

Over a dozen states now offer elementary mathematics specialist certification, and many universities offer graduate degree programs for those wishing to specialize in elementary mathematics education. As with other courses and programs for elementary teachers, mathematicians and mathematics educators have opportunities to work together to develop and teach courses for EMS.

Early Childhood Teachers

Younger children are naturally inquisitive and can be powerful and motivated mathematical learners, who are genuinely interested in exploring mathematical ideas. Currently, there are large disparities in the mathematical abilities of young children. These are linked to socioeconomic status and are larger in the United States than in some other countries. According to the National Research Council report *Mathematics Learning in Early Childhood*, "there is mounting evidence that high-quality preschool can help ameliorate inequities in educational opportunity and begin to address achievement gaps," but "many in the early childhood workforce are not aware of what young children are capable of in mathematics and may not recognize their potential to learn mathematics." Early childhood teachers sometimes hold a variety of beliefs that are not supported by current research.

---

[16]In general, a math specialist's roles and responsibilities are not analogous to those of a reading specialist.

[17]Examples include the Vermont Mathematics Initiative (a Math Science Partnership), see *Teaching Teachers Mathematics*, Mathematical Sciences Research Institute, 2009, pp. 36–38. A 3-year randomized study in Virginia found that specialists' coaching of teachers had a significant positive effect on student achievement in grades 3–5. The specialists studied completed a mathematics program designed by the Virginia Mathematics and Science Coalition (also a Math Science Partnership) and the findings should not be generalized to specialists with less expertise. See Campbell & Malkus, "The Impact of Elementary Mathematics Coaches on Student Achievement," *Elementary School Journal*, 2011.

[18]*Standards for Elementary Mathematics Specialists: A Reference for Teacher Credentialing and Degree Programs*, Association of Mathematics Teacher Educators, 2009.

These may include "Young children are not ready for mathematics education" or "Computers are inappropriate for the teaching and learning of mathematics."[19]
Mathematics Learning in Early Childhood states:

> Coursework and practicum requirements for early childhood educators should be changed to reflect an increased emphasis on children's mathematics as described in the report. These changes should also be made and enforced by early childhood organizations that oversee credentialing, accreditation, and recognition of teacher professional development programs.

Designers of preparation programs are advised to review their coursework in early childhood mathematics and to prepare teachers in the following areas:

- mathematical concepts and children's mathematical development;

- curricula available for teaching mathematics to young children;

- assessment of young children's mathematical skills and thinking and how to use assessments to inform and improve instructional practice; and

- opportunities to explore and discuss their attitudes and beliefs about mathematics and the effects of those beliefs on their teaching.[20]

Coursework to address these topics satisfactorily will take 6 to 9 semester-hours.

Teachers of Special Populations

The Council for Exceptional Children distinguishes between the roles of teachers "in the core academic subjects" versus other roles that special education teachers play (e.g., co-teaching, helping to design individualized education programs). Similarly, teachers who work with students who are English Language Learners (ELLs) may be teaching mathematics or may be working with students in other capacities (such as developing their language skills and helping them adapt socially). Special education teachers and ELL teachers who have direct responsibility for teaching mathematics (a core academic subject) should have the same level of mathematical knowledge as general education teachers in the subject.

MET II's recommendations for preparation and professional development apply to special education teachers, teachers of ELL students, and any other teacher with direct responsibility for teaching mathematics.

---

[19]These examples come from Lee & Ginsburg, "Early Childhood Teachers' Misconceptions about Mathematics Education for Young Children in the United States," *Australasian Journal of Early Childhood*, 2009. This article summarizes research in this area and discusses possible sources of such beliefs.

[20]See *Mathematics Learning in Early Childhood*, National Research Council, 2009, pp. 341–343.

CHAPTER 5

Middle Grades Teachers

What mathematics and statistics should future middle grades teachers study to prepare for their careers? What kinds of mathematics coursework and programs will prepare middle grades teachers for teaching mathematics? What professional development experiences will both develop and sustain high quality mathematics teaching in the middle grades? How can mathematicians make valuable contributions to these endeavors? These questions are the topics of this chapter. Coursework in mathematical pedagogy is assumed to be part of a preparation program, but is not discussed in detail.

In this chapter, the term "middle grades teacher" is defined as a teacher who teaches mathematics in grade 6, 7, or 8.[1] The chapter addresses the mathematical knowledge that a middle grades teacher needs to teach, and teach well, the mathematics described for grades 6–8 in the Common Core State Standards for Mathematics (CCSS).

It is important to note that there are distinctions among state requirements for certification to teach mathematics at various grade levels and the requirements found in different teacher preparation programs. Currently 46 states and the District of Columbia provide a license, certificate, or endorsement specific to middle grades. In all but two cases, grade 5 is one of the grades included in the credential.[2]

Many institutions of higher education that prepare teachers do not offer a program specifically and exclusively designed for middle grades teachers of mathematics.[3] Indeed, the majority of middle grades teachers are likely prepared in a program designed as preparation to teach all academic subjects in grades K–8 or in a program to teach mathematics in grades 7–12 or 6–12. Moreover, programs that do offer specific preparation for middle grades often lead to multi-subject certification (such as a certificate to teach mathematics and science), making it challenging for future teachers to take all the mathematics recommended by this report.

Regardless of where middle grades teachers are prepared and how they are certified, it is critical that they have the opportunity to understand the mathematics

---

Note that the MET II web resources at www.cbmsweb.org give URLs for the CCSS, the Progressions for the CCSS, and other relevant information.

[1]As noted in Chapter 3, "Although elementary certification in most states is still a K–6 and, in some states, a K–8 certification, state education departments and accreditation associations are urged to require all grades 5–8 teachers of mathematics to satisfy the 24-hour requirement recommended by this report." Chapters 4 and 5 allow for a period of transition.

[2]See the listing at the Association for Middle Level Education web site.

[3]See, e.g., Tatto & Senk, "The Mathematics Education of Future Primary and Secondary Teachers: Methods and Findings from the Teacher Education and Development Study in Mathematics," *Journal of Mathematics Teacher Education*, 2011, p. 127; *Report of the 2000 National Survey of Mathematics and Science Education*, Horizon Research, p. 16.

in the middle grades from a teacher's perspective. Over time, middle grades mathematics has become more challenging, and the Common Core State Standards outline a significant change in its content, as well as its depth and breadth, for these grades. Many current middle grades teachers, particularly those who teach in grade 6, have elementary certification. Thus, long after it becomes commonplace for future sixth-grade teachers (and many fifth-grade teachers) to earn certification through a middle grades mathematics teacher preparation program, there will be a significant need for content-based professional development opportunities for teachers of mathematics in grades 5 through 8.

Essential Grades 6–8 Ideas for Teachers

This section uses the CCSS as a framework for describing the mathematics that middle grades teachers, both prospective and practicing, should study and know. The CCSS standards for mathematics content are organized into clusters of related standards and the clusters are organized into mathematical domains, which span multiple grade levels (see Appendix B). Brief descriptions of how the mathematics of each domain progresses across grade levels and is connected within or across grades to standards in other domains appear in the Progressions for the CCSS (see the web resources associated with this report).

Because middle grades teachers receive their students from elementary school and prepare them for high school, college courses and professional development opportunities for middle grades teachers should also attend to how the mathematical ideas of the middle grades connect with ideas and topics of elementary school and high school. Thus, courses and professional development will need to devote time to ideas within the K–5 and high school domains (see Chapters 4 and 6).

In this section, essential mathematical ideas are listed for each 6–8 CCSS domain. Teachers need to know these ideas well, but the listings are not intended to be comprehensive. Instructors are encouraged to refer to the CCSS, related progressions, and other references given in the web resources for further details and discussion.

Each list of essential ideas for a domain is followed by a list of related activities that illustrate the ideas and could be used in teacher preparation or professional development.

A given activity may provide opportunities to demonstrate or develop various kinds of expertise described by one or more of the CCSS standards for mathematical practice. These are indicated by the number and heading of the associated standard. For example, "MP 1 Make sense of problems and persevere in solving them" indicates that expertise connected with the first Standard for Mathematical Practice might be used. (The full text for all eight Standards for Mathematical Practice appears as Appendix C of this report.) Note that although a particular activity might provide opportunities to use or increase expertise, instructors should expect to foster engagement in these opportunities. Also, instructors might periodically ask middle grades teachers to review and reflect on the Standards for Mathematical Practice, encouraging teachers to become more familiar with these standards and how they may be achieved in middle grades mathematics.

Ratio and Proportional Relationships (Grades 6–7)[4]

- Reasoning about how quantities vary together in a proportional relationship, using tables, double number lines, and tape diagrams as supports.
- Distinguishing proportional relationships from other relationships, such as additive relationships and inversely proportional relationships.
- Using unit rates to solve problems and to formulate equations for proportional relationships.
- Recognizing that unit rates make connections with prior learning by connecting ratios to fractions.
- Viewing the concept of proportional relationship as an intellectual precursor and key example of a linear relationship.

Illustrative activities:

(1) Examine different ways to solve proportion problems with tables, double number lines, and tape diagrams. Examine common errors students make when solving problems involving ratio and proportion.

MP 2 Reason quantitatively and abstractly.
MP 3 Construct viable arguments and critique the reasoning of others.
MP 5 Use appropriate tools strategically by reasoning with visual models.

(2) Compare and contrast different ways to find values in proportional relationships and in inversely proportional relationships. For example, explain why linear interpolation can be used with proportional relationships but not with inversely proportional relationships.

MP 3 Construct viable arguments and critique the reasoning of others.
MP 4 Model with mathematics.
MP 7 Look for and make use of structure.

The Number System (Grades 6–8)

- Understanding and explaining methods of calculating products and quotients of fraction, by using area models, tape diagrams, and double number lines, and by reading relationships of quantities from equations.
- Using properties of operations (the CCSS term for the field axioms) to explain operations with rational numbers (including negative integers).
- Examining the concepts of greatest common factor and least common multiple.
- Using the standard U.S. division algorithm to explain why decimal expansions of fractions eventually repeat and showing how decimals that eventually repeat can be expressed as fractions.
- Explaining why irrational numbers are needed and how the number system expands from rational to real numbers.

---

[4]See the Ratio and Proportion Progression for further details, including examples of double number lines and tape diagrams, and discussion of unit rates. In the CCSS, "fractions" refers to non-negative rational numbers in grades 3–5. Note that distinctions made in the CCSS between fractions, ratios, and rates may be unfamiliar to teachers.

Illustrative activities:

(1) Solve fraction division problems using the Group Size Unknown (sharing) perspective and the Number of Groups Unknown (measurement) perspective on division with tape diagrams and double number lines.[5] Use these approaches, as well as the connection between multiplication and division (division can be viewed as multiplication with an unknown factor), to develop rationales for methods of computing quotients of fractions.

MP 2 Reason abstractly and quantitatively.
MP 3 Construct viable arguments and critique the reasoning of others.
MP 5 Use appropriate tools strategically.

(2) Explain why rules for adding and subtracting with negative numbers make sense by using properties of operations (e.g., commutativity and associativity of addition and additive inverses) and the connection between addition and subtraction (subtraction can be viewed as finding an unknown addend). Similarly, explain why rules for multiplying and dividing with negative numbers make sense.

MP 3 Construct viable arguments and critique the reasoning of others.
MP 8 Look for and express regularity in repeated reasoning.

(3) Use the standard U.S. division algorithm to explain why the length of the string of repeating digits in the decimal expansion of a fraction is at most 1 less than the denominator. Explain why $0.999\ldots = 1$ in multiple ways.[6]

MP 7 Look for and make use of structure.
MP 8 Look for and express regularity in repeated reasoning.

(4) Prove that there is no rational number whose square is 2.

MP 3 Construct viable arguments and critique the reasoning of others.
MP 7 Look for and make use of structure.

Expressions and Equations (Grades 6–8)

- Viewing numerical and algebraic expressions as "calculation recipes," describing them in words, parsing them into their component parts, and interpreting the components in terms of a context.
- Examining lines of reasoning used to solve equations and systems of equations.
- Viewing proportional relationships and arithmetic sequences as special cases of linear relationships. Reasoning about similar triangles to develop the equation $y = mx + b$ for (non-vertical) lines.

---

[5]For descriptions of multiplication and division problem types, see the CCSS, p. 89 or the Operations and Algebraic Thinking Progression.

[6]"I don't think it's equal because I think that would be confusing to kids to say that 99 cents can be rounded up to a dollar" and other examples of conceptions that teachers may hold about this equation are given in Yopp et al., "Why It is Important for In-service Elementary Mathematics Teachers to Understand the Equality $.999\ldots = 1$," *Journal of Mathematical Behavior*, 2008. Note that undergraduates may use decimal notation in ways that suggest notions of nonstandard analysis, see Ely, "Nonstandard Student Conceptions About Infinitesimals," *Journal for Research in Mathematics Education*, 2010.

Illustrative activities:

(1) Use tape diagrams as tools in formulating and solving problems and connect the solution strategy to standard algebraic techniques.

MP 1 Make sense of problems and persevere in solving them.
MP 4 Model with mathematics.
MP 5 Use appropriate tools strategically.
MP 7 Look for and make use of structure.

(2) Reason about entries of sequences. In particular, determine that and
explain why arithmetic sequences are described by formulas of the form
$y = mx + b$.

MP 4 Model with mathematics.
MP 7 Look for and make use of structure.
MP 8 Look for and express regularity in repeated reasoning.

(3) Examine how different types of equations are used for different purposes.
(Some equations show the result of a calculation, some equations are to
be solved when solving a problem, some equations are used to describe
how two quantities vary together, and some equations express identities,
such as the distributive property.)

MP 2 Reason abstractly and quantitatively.

Functions (Grade 8)

- Examining and reasoning about functional relationships represented using
tables, graphs, equations, and descriptions of functions in words. In particular, examining how the way two quantities change together is reflected
in a table, graph, and equation.
- Examining the patterns of change in proportional, linear, inversely proportional, quadratic, and exponential functions, and the types of real-world
relationships these functions can model.

Illustrative activities:

(1) Given a graph, tell a story that fits with the graph. Given a story, create
a graph that fits with the story.

MP 2 Reason abstractly and quantitatively.
MP 3 Construct viable arguments and critique the reasoning of others.
MP 4 Model with mathematics.

(2) Compare and contrast equations, graphs, patterns of change, and types of
situations modeled by different relationships. For example, compare and
contrast inversely proportional relationships and linear relationships that
have graphs with negative slopes. Compare and contrast linear relationships and exponential relationships (including arithmetic sequences and
geometric sequences, e.g., contrast repeatedly adding 5 with repeatedly
multiplying by 5).

MP 2 Reason abstractly and quantitatively.
MP 4 Model with mathematics.
MP 7 Look for and make use of structure.

Geometry (Grades 6–8)

- Deriving area formulas such as the formulas for areas of triangles and parallelograms, considering the different height–base cases (including the "very oblique" case where "the height is not directly over the base").
- Explaining why the Pythagorean Theorem is valid in multiple ways. Applying the converse of the Pythagorean Theorem.
- Informally explaining and proving theorems about angles; solving problems about angle relationships.
- Examining dilations, translations, rotations, and reflections, and combinations of these.
- Understanding congruence in terms of translations, rotations, and reflections; and similarity in terms of translations, rotations, reflections, and dilations; solving problems involving congruence and similarity in multiple ways.

Illustrative activities:

(1) Find and explain angle relationships, e.g., the sum of the angles in a 5-pointed star drawn with 10 line segments or the sum of the exterior angles of a shape. Illustrate, informally demonstrate, and prove that the sum of the angles in a triangle is always 180 degrees. Discuss the distinction between an informal demonstration and a proof, as well as ways in which a demonstration can suggest steps in a proof (e.g., tearing off corners and putting them together may suggest the strategy of drawing an auxiliary line).

MP 1 Make sense of problems and persevere in solving them.
MP 3 Construct viable arguments and critique the reasoning of others.
MP 4 Model with mathematics.
MP 7 Look for and make use of structure.

(2) Examine how area and volume change between similar shapes.

MP 5 Use appropriate tools strategically.
MP 7 Look for and make use of structure.
MP 8 Look for and express regularity in repeated reasoning.

Statistics and Probability (Grades 6–8)

- Understanding various ways to summarize, describe, and compare distributions of numerical data in terms of shape, center, and spread.
- Calculating theoretical and experimental probabilities of simple and compound events, and understanding why their values may differ for a given event in a particular experimental situation.
- Developing an understanding of statistical variability and its sources, and the role of randomness in statistical inference.
- Exploring relationships between two variables by studying patterns in bivariate data.

Illustrative activities:

(1) Investigate patterns in repeated random samples or probability experiments to develop a robust understanding of "random."

MP 4 Model with mathematics.
MP 5 Use appropriate tools strategically.

(2) Compare and contrast various measures of center and spread as well as means of calculating them.

MP 4 Model with mathematics.
MP 7 Look for and make use of structure.

(3) Identify sources of variability in data, and draw inferences from analyses of the data.

MP 2 Reason abstractly and quantitatively.
MP 6 Attend to precision.

The Preparation and Professional Development of Middle Grades Teachers

Because the middle grades are "in the middle," it is critical that middle grades teachers be aware of the mathematics that students will study before and after the middle grades. This has significant implications for the preparation of and professional development of middle grades teachers. Middle grades teachers need to be well versed in the mathematics described in Chapter 4, particularly in the domains pertaining to whole numbers and fractions.[7] Moreover, middle grades teachers need to know how the topics they teach are connected to later topics so that they can introduce ideas and representations that will facilitate students' learning of mathematics in high school and beyond. For instance, prospective and practicing middle grades teachers need to be aware of representations, be they drawings, tape diagrams, number lines,[8] or physical models, used in the earlier grades and how those representations may lend themselves to establishing and extending mathematical ideas into the middle grades.[9] For instance, in grades 3–5, area models may be used to represent a product of two fractions, but linear models such as tape diagrams and double number lines are important in the middle grades because they are more readily connected to representations of numbers on the number line and the coordinate plane. Area models may not lead students to successive partitions, which is necessary when thinking about how to partition the interval from 0 to 1 into sixths (partition in half, then partition each half in thirds; or vice versa). On the other hand, area models are used in estimating area under a curve in calculus, and ratio and proportion are intellectual precursors for linear functions.

These examples illustrate the need for middle grades teachers to specialize in mathematics and why their preparation should specifically address mathematics

---

[7]It is important for middle grades teachers to have an elementary teacher's perspective on this content because they may need to provide support and instruction for students who have not yet achieved proficiency.

[8]Note that the CCSS use the term "number line diagram" instead of "number line."

[9]Examples of these representations occur in the Progressions for the CCSS.

relevant for teaching grades 5–8. Although this chapter focuses on the mathematics taught in grades 6–8, grade 5 is included here for two reasons. First, teachers receive students from grade 5, thus need to understand the mathematics of grade 5. Second, in most states, middle grades certification includes certification to teach grade 5 and, as recommended in Chapter 3, there should be a transition to the expectation that grade 5 teachers have middle grades certification.

Many, if not most, middle schools offer courses, such as algebra and geometry for high school credit. A middle grades teacher of such courses should have further preparation that goes beyond the recommendations of this chapter, or its professional development equivalent, and be prepared to work closely with high school colleagues in developing appropriate transitions between middle and high school mathematics.

Programs for Prospective Teachers

The mathematics outlined by the Common Core State Standards for grades 6–8 is intellectually challenging and middle grades teachers will require substantial preparation in order to teach it. Initial study of the mathematics for teaching middle grades requires at least 24 semester-hours. At least 15 of these semester-hours should consist of mathematics courses designed specifically for future middle grades teachers that address the essential ideas described in the previous section and in Chapter 4. The remaining 9 semester-hours should include courses that will strengthen prospective teachers' knowledge of mathematics and broaden their understanding of mathematical connections between one grade band and the next, connections between elementary and middle grades as well as between middle grades and high school. This second type of coursework should be carefully selected from mathematics and statistics department offerings. In no case should a course at or below the level of precalculus be considered part of these 24 semester-hours.

Mathematics and statistics courses designed for future middle grades teachers. First and foremost, future teachers need courses that allow them to delve into the mathematics of the middle grades while engaging in mathematical practice as described by the CCSS. The instructors of these courses should model good pedagogy. The courses should be taught with the understanding that the course-takers are future teachers so efforts should be made to connect the mathematics they are learning to mathematics they will teach and challenges they will face when teaching it. These courses should be designed specifically for future middle grades teachers.

Essential ideas that teachers should study in depth and from a teacher's perspective are outlined in the preceding section of this chapter and in Chapter 4. These ideas can be studied in the courses described below. Note that topics which are names of domains in the CCSS (e.g., "ratio and proportional relationships") refer to clusters of ideas described in the corresponding domain and progression for the CCSS.

> *Number and operations* (6 semester-hours). Number and operations in base ten, fractions, addition, subtraction, multiplication, and division with whole numbers, decimals, fractions, and negative numbers. Possible additional topics are irrational numbers or arithmetic in bases other than ten.

Depending on course configuration, some of the topics listed below in the algebra and number theory course might be addressed in a number and operations course.

Geometry and measurement (3 semester-hours). Perimeter, area, surface area, volume, and angle; geometric shapes, transformations, dilations, symmetry, congruence, similarity; and the Pythagorean Theorem and its converse.

Algebra and number theory (3 semester-hours). Expressions and equations, ratio and proportional relationships (and inversely proportional relationships), arithmetic and geometric sequences, functions (linear, quadratic, and exponential), factors and multiples (including greatest common factor and least common multiple), prime numbers and the Fundamental Theorem of Arithmetic, divisibility tests, rational versus irrational numbers. Additional possible topics for teachers who have already studied the above topics in depth and from a teacher's perspective are: polynomial algebra, the division algorithm and the Euclidean algorithm, modular arithmetic.

Statistics and probability (3 semester-hours). Describing and comparing data distributions for both categorical and numerical data, exploring bivariate relationships, exploring elementary probability, and using random sampling as a basis for informal inference.

A necessary prerequisite for the statistics and probability course for middle level teachers is a modern introductory statistics course emphasizing data collection and analysis. This background will allow the course designed for middle grades teachers to emphasize active learning with appropriate hands-on devices and technology while probing deeply into the topics taught in the middle grades, all built around seeing statistics as a four-step investigative process involving question development, data production, data analysis and contextual conclusions.[10]

Other mathematics and statistics courses. This second type of coursework should be carefully selected from mathematics or statistics department offerings that are both useful for and accessible to undergraduates in the institution's middle level education program. It should include:

Introductory statistics. As noted above, this is a recommended prerequisite for the statistics and probability course designed for teachers. The introductory course should have a modern technology-based emphasis with topics that include basic principles of designing a statistical study, data analysis for both categorical and numerical data, and inferential reasoning, much as in the introductory statistics and probability course for high school teachers described in Chapter 6. (In some departments, this might be the same course.)

---

[10]See the *Guidelines for Assessment and Instruction in Statistics Education (GAISE) Report: A PreK–12 Curriculum Framework* of the American Statistical Association.

Other courses might include:

Calculus. Although many institutions require future middle grades mathematics teachers to take the standard first-semester calculus course for engineers, a "concepts of calculus" course might be more useful for those who will be middle level teachers. Such a course could include careful study of the concepts underlying standard topics of calculus (e.g., sequences, series, functions, limits, continuity, differentiation, optimization, curve sketching, anti-differentiation, areas of plane regions, lengths of plane curves, areas of surfaces of revolution, and volumes of solids).

Number theory. One possibility is a course that focuses on basic number theory results needed to understand the number theoretic RSA cryptography algorithm. As the number theory results are developed, connections to middle level curricula could be emphasized. Proofs should be carefully selected so that they are particularly relevant and accessible to middle level teachers.

Discrete mathematics. This can offer teachers an opportunity to explore in depth many of the topics they will teach. Possible discrete mathematics topics introduced in this course could include social decision-making, vertex-edge graph theory, counting techniques, matrix models, and the mathematics of iteration. The unifying themes for these topics should be mathematical modeling, the use of technology, algorithmic thinking, recursive thinking, decision-making, and mathematical induction as a way of knowing.

History of mathematics. A history of mathematics course can provide middle grades teachers with an understanding of the background and historical development of many topics in middle grades (see examples in Chapter 6). Past applications of topics can illustrate their uses in modeling, thus sometimes their historical significance.

Modeling. A substantive mathematical modeling course can provide prospective teachers with understanding of the ways in which mathematics and statistics can be applied.

Methods courses. In addition to the mathematics courses described above, prospective middle grades teachers should take two methods courses that address the teaching and learning of mathematics in grades 5–8. At some institutions, it may be possible, and even desirable, to create hybrid courses that integrate the study of mathematics and pedagogy. In these situations, it is still imperative that future teachers complete the equivalent of 24 semester-hours of mathematics.

PROFESSIONAL DEVELOPMENT FOR PRACTICING TEACHERS

Teachers of middle grades students must be able to build on their students' earlier mathematics learning and develop a broad set of new understandings and skills to help students meet these more sophisticated mathematical goals. Teaching middle grades mathematics requires preparation *different from* preparation for teaching high school mathematics. Once they begin teaching, middle school teachers need continuing opportunities to deepen and strengthen their mathematical knowledge for teaching, particularly as they engage with students and develop better understanding of their thinking.

Although professional development experiences for middle grades teachers may take a variety of forms, the central focus should be providing opportunities to deepen and strengthen mathematical knowledge in the domains of the CCSS. Many current teachers prepared before the era of the Common Core State Standards will need opportunities to study and learn mathematics and statistics that they have not previously taught. Prior to the CCSS, mathematics in grades 6–8 focused heavily on work with rational numbers (including computational fluency), as well as development of proficiency with geometric measurement (area, surface area, volume) and readiness for algebra (introduction to negative integers, expressions, and equations). In the CCSS, many of these concepts are developed earlier. The shifts in curriculum focus represented by the CCSS (e.g., increased attention to algebra) and the new topics (e.g., transformational approach to congruence), present challenges for many middle grades teachers and underscore the need for professional development.

Professional development for teachers may take many different forms. A group of teachers might work together in a professional learning community, focusing deeply on one topic for a period of time. For example, sixth-grade teachers within a school (or across several schools) might spend a term designing, teaching, and analyzing lessons on expressions and equations using a lesson study format.[11] Or a group of teachers who teach grades 6, 7, and 8 at one school might meet regularly to study how a topic such as proportional reasoning develops across grade levels. A group of teachers might watch demonstration lessons and then meet to discuss the lessons, plan additional lessons, and study the mathematics of the lessons. Math teachers' circles and immersion experiences are additional options.[12] Teachers might also complete mathematics courses specifically designed as part of a graduate program for middle grades mathematics teachers.

Regardless of the format, professional development should engage teachers in mathematics. It should include opportunities for discussing student learning of this mathematics, common student misconceptions, the ways that ideas in the CCSS are related to and build upon one another, and the most useful representations, tools (electronic and otherwise), and strategies for teaching this mathematics to students. However it is organized, as discussed in Chapter 2, the best professional development is ongoing, directly related to the work of teaching mathematics, and focused on mathematical ideas.

Mathematicians and mathematics educators in higher education play an important role in helping to organize, facilitate, or contribute to the professional development of middle school teachers. In doing so, they also have opportunities to think

---

[11]Lesson study is a process in which teachers jointly plan, observe, analyze, and refine actual classroom lessons.

[12]Math teachers' circles and immersion experiences focus primarily on giving teachers an experience to be learners and doers of mathematics. See Chapter 6 and the web resources for further information and examples.

about undergraduate mathematics teaching and connections between college-level
mathematics and K–12 education.

Engaging in mathematical practice. The CCSS standards for mathematical practice describe features of mathematical expertise that learners (including
prospective and practicing teachers) develop as they do mathematics. Although
these are often discussed separately from mathematical topics, the two should be
viewed as inseparable. That is, when doing mathematics, one is engaging in mathematical practice. These features of mathematical practice must be made explicit in
preparation and professional development programs. Teachers need to know what
they are, and to be able to identify instances in their own work on a particular problem and in children's work, and to be able to think explicitly about when, where,
and how these types of expertise would occur in middle grades mathematics.

Using technology and other tools strategically. The tools available for
teaching middle grades mathematics include interactive whiteboards and tablets,
mathematics-specific technology such as virtual manipulatives, dynamic geometry
software, graphing calculators and programs, and an ever-expanding collection of
applets, apps, web sites, and multimedia materials. It is essential that prospective and practicing teachers have opportunities to use such tools as they explore
mathematical ideas in order to enhance their mathematical thinking, expand the
repertoire of technological tools with which they are proficient, and develop an
awareness of the limitations of technology. Teachers learn to use technology as a
computational tool to perform a calculation or produce a graph or table in order
to use the result as input to analyze a mathematical situation. They should also
learn to use technology as a problem solving tool, or to conduct an investigation by
taking a deliberate mathematical action, observing the consequences, and reflecting
on the mathematical implications of the consequences. Teachers must have opportunities to engage in the use of a variety of technological tools, including those
designed for mathematics and for teaching mathematics, to explore and deepen
their understanding of mathematics, even if these tools are not the same ones they
will eventually use with students.

Technology is one of many tools available for learning and teaching mathematics. Others are traditional tools of teaching such as blackboards[13] and geometric
models, and newer ones such as patty paper, mirrors, and tangrams.[14] Teachers
need to develop the ability to critically evaluate the affordances and limitations
of a given tool, both for their own learning and to support the learning of their
students. In mathematics courses for teachers, instructors should model successful
ways of using tools for teaching, and provide opportunities to discuss mathematical
issues that arise in their use.

Challenges in the Education of Middle Grades Teachers

Prospective middle school teachers enter teacher preparation programs with
their own views about what it means to know and do mathematics and how it is
learned. Because they have chosen to become mathematics teachers, they are likely

---

[13]See, e.g., discussion of the use and organization of the blackboard in Lewis, *Lesson Study*,
Research for Better Schools, 2002, pp. 97–98.

[14]See Kidwell et al., *Tools for Teaching Mathematics in the United States, 1800–2000*, Johns
Hopkins University Press, 2008.

to be confident in their abilities to learn mathematics. In this sense, these teacher candidates are more likely to be more similar to pre-service high school teachers than to pre-service elementary teachers. However, their perspective on what it means to know mathematics may be based on their own success in learning facts and procedures rather than on understanding the underlying concepts upon which the procedures are based.

Pre-service middle grades teachers may not be familiar with all of the expectations outlined in the CCSS for middle school students. Thus, they may question the need to learn things in their teacher preparation program that were not part of their own middle grades mathematics.

Although many states offer a distinct certification for middle school teachers (e.g., grades 5–9), other states award middle school endorsements or licenses to teachers together with elementary certification or as part of high school certification (e.g., a teacher is certified to teach grades 7–12). In some states, certified teachers can obtain middle school certification in mathematics by taking and passing an exam such as the Middle School Praxis without taking additional mathematics coursework. Thus, teachers of middle school mathematics are diverse in their mathematical preparation—some have studied the same mathematics as high school teachers; others have completed a few mathematics courses beyond requirements for elementary teachers. Therefore, professional development for middle school mathematics teachers must acknowledge the diversity of background knowledge, both mathematical and pedagogical, that teachers at this level, within a school, district, or state, may have. In any case, the focus of professional development should be on understanding the mathematics outlined in the CCSS for grades 5–8 and instructional strategies to support students in learning it.

Teachers of Special Populations

The Council for Exceptional Children distinguishes between the roles of teachers "in the core academic subjects" versus other roles that special education teachers play (e.g., co-teaching, helping to design individualized education programs). Similarly, teachers who work with students who are English Language Learners (ELLs) may be teaching mathematics or may be working with students in other capacities (such as developing their language skills and helping them adapt socially). Special education teachers and ELL teachers who have direct responsibility for teaching mathematics (a core academic subject) should have the same level of mathematical knowledge as general education teachers in the subject.

MET II's recommendations for preparation and professional development apply to special education teachers, teachers of ELL students, and any other teacher with direct responsibility for teaching mathematics.

CHAPTER 6

High School Teachers

What mathematics should prospective high school teachers study to prepare for their careers? What kinds of coursework and programs will prepare high school teachers for teaching mathematics? What sorts of professional development experiences will develop and sustain high quality mathematics teaching in high school? How can mathematicians make valuable contributions to these endeavors? These questions are the topics of this chapter. Coursework in mathematical pedagogy is assumed to be part of a preparation program, but is not discussed in detail.

In *Elementary Mathematics from an Advanced Standpoint*, Felix Klein described what he called the double discontinuity experienced by prospective high school teachers:

> The young university student [was] confronted with problems that did not suggest ... the things with which he had been concerned at school. When, after finishing his course of study, he became a teacher ... he was scarcely able to discern any connection between his task and his university mathematics.[1]

The double discontinuity consists of the jolt experienced by the high school student moving from high school to university mathematics, followed by the second jolt moving from the mathematics major to teaching high school. This discontinuity still exists today. Many high school students, even those who are successful in their mathematics courses, graduate with a view of mathematics as a static body of knowledge and skills, full of special-purpose tools and methods that are used to solve small classes of problems. Missing is the overall coherence and parsimony of the discipline and the beautiful simplicity of a subject in which a small number of ideas can be used to build intricate and textured edifices of interconnected results.

As noted in the first MET report, analyses by Ed Begle in the 1970s and David Monk in the 1990s suggest that the set of upper-division courses typical of a mathematics major have minimal impact on the quality of a teacher's instruction, as measured by student performance.[2] This is not to say that subject matter of those courses is not valuable for teachers. However, it may be that the choice of topics and the way they are developed is not helpful. For example, teachers might emerge from a course on Galois theory without having seen its connection with the quadratic formula. In this regard, Hung-Hsi Wu makes the following recommendation for the preparation of high school teachers:

Note that the MET II web resources at www.cbmsweb.org give URLs for the CCSS, the Progressions for the CCSS, and other relevant information.

[1]Translation of the third edition, Macmillan, 1932, p. 1.

[2]This line of research and its limitations are discussed in more detail in Chapter 2.

> In contrast with the normal courses that are relentlessly "forward-looking" (i.e., the far-better-things-to-come in graduate courses), considerable time should be devoted to "looking back."[3]

Thus, one theme of this chapter is that the mathematical topics in courses for prospective high school teachers and in professional development for practicing teachers should be tailored to the work of teaching, examining connections between middle grades and high school mathematics as well as those between high school and college.

A second theme concerns not the topics studied but the practice of mathematics. The National Academy report *Adding It Up* defines five strands of mathematical proficiency: conceptual understanding, procedural fluency, strategic competence, adaptive reasoning, and productive disposition. High school teachers have the responsibility of building on their students' mathematical experiences in previous grades to give them a sturdy proficiency composed of all five strands interwoven. The Common Core State Standards for Mathematical Practice describe how this proficiency might look in various mathematical situations and specific examples are given in the Progressions for the CCSS.

To achieve this result, teachers need opportunities for the full range of mathematical experience themselves: struggling with hard problems, discovering their own solutions, reasoning mathematically, modeling with mathematics, and developing mathematical habits of mind. Thus, in addition to describing topics, this chapter describes varieties of mathematical experience for teacher preparation and professional development.

Some topics and experiences will occur during preparation, others will occur in professional development over the course of teachers' careers. This chapter describes:

- Essentials in the mathematical preparation of high school teachers.

- Important additional mathematics content that can be learned in undergraduate electives or in professional development programs for practicing teachers.

- Essential mathematical experiences for practicing teachers.

With regard to topics, in all three categories, teachers might take standard courses for mathematics majors. This chapter describes ways in which such courses can be adjusted to better connect with the mathematics of high school. Here "the mathematics of high school" does not mean simply the syllabus of high school mathematics, the list of topics in a typical high school text. Rather it is the structure of mathematical ideas from which that syllabus is derived.

Much of this structure is absent from current courses that prospective teachers often take, either in high school or in college. For example, the method of completing the square may or may not be present, as a pure technique, in a high school algebra course. However, there is more to know about completing the square than how to carry out the technique: it reduces every single-variable quadratic equation

---

[3]"On the Education of Mathematics Majors" in *Contemporary Issues in Mathematics Education*, Mathematical Sciences Research Institute, 1999, p. 13.

to an equation of the form $x^2 = k$, and thereby leads to a general formula for the solution of a single-variable quadratic equation; it generalizes to a method of eliminating the next to highest order term in higher order equations; it allows one to translate the graph of every quadratic function so that its vertex is at the origin, and thereby allows one to show that all such graphs are similar; and it provides an important step in simplifying quadratic equations in two variables, leading to a classification of the graphs of such equations. The treatment of completing the square in high school is often the merest decoration on this body of knowledge, and a university mathematics major might hear no more of the matter. A prospective teacher who sees some of these connections is better prepared to teach completing the square in a manner consistent with the Standards for Mathematical Practice in the Common Core.

This report recommends that the mathematics courses taken by prospective high school teachers include at least a three-course calculus sequence, an introductory statistics course, an introductory linear algebra course, and 18 additional semester-hours of advanced mathematics, including 9 semester-hours explicitly focused on high school mathematics from an advanced standpoint. It is desirable to have a further 9 semester-hours of mathematics; the appendix for this chapter gives suggestions for a short and a long mathematics course sequence exemplifying these recommendations. A full program would also include all education courses required for certification, which are not described in this report. It is recommended that the methods courses required for certification focus on instructional strategies for high school mathematics rather than generic instructional methods.

Whatever the length of the program, the recommendations described here, particularly the 9 semester-hours of coursework designed for prospective teachers, are ambitious and will take years to achieve. They are, however, what is needed. Institutions that serve only a few prospective teachers per year may be unable to offer many courses with its prospective teachers as the sole audience. These programs may need to consider innovative solutions, similar to the regional centers recommended by the National Task Force on Teacher Education in Physics (see the web resources for examples).[4] Furthermore, courses designed for prospective high school teachers can also serve the needs of other mathematics majors, of prospective middle school teachers if the department has a dedicated program for them, and, if offered at convenient times and locations, of practicing teachers seeking further education.

Essentials in Mathematical Preparation

A primary goal of a mathematics major program is the development of mathematical reasoning skills. This may seem like a truism to higher education mathematics faculty, to whom reasoning is second nature. But precisely because it is second nature, it is often not made explicit in undergraduate mathematics courses. A mathematician may use reasoning by continuity to come to a conjecture, or delay the numerical evaluation of a calculation in order to see its structure and create

---

[4]See Recommendation 13 of *National Task Force on Teacher Education in Physics: Report Synopsis*, American Association of Physics Teachers, the American Physical Society, & the American Institute of Physics, 2010.

a general formula, but what college students see is often the end result of this thinking, with no idea about how it was conceived.

Reasoning from known results and definitions supports retention of knowledge in a mathematical domain by giving it structure and connecting new knowledge to prior knowledge. This kind of reasoning is useful in most careers, even if domain knowledge is forgotten. It is especially important for teachers, because a careful look at the mathematics that is taught in high school reveals that it is often developed as a collection of unrelated facts that are not always justified or precisely formulated.[5] Hence, many incoming undergraduates are not used to seeing the discipline as a coherent body of connected results derived from a parsimonious collection of assumptions and definitions. One necessary ingredient to breaking this cycle is the next generation of teachers, who must have a coherent view of the structure of mathematics in order to develop reasoning skills in their students. Thus, this report recommends that when courses include prospective teachers, instructors pay careful attention to building and guiding mathematical reasoning—generalizing, finding common structures in theorems and proofs, seeing how a subject develops through a sequence of theorems, and forming connections between seemingly unrelated concepts. At the heart of mathematical reasoning is asking the right questions. As George Pólya liked to tell his classes, "It is easy to teach students the right answers; the challenge is to teach students to ask the right questions."

To teach mathematical reasoning requires a classroom where learners are active participants in developing the mathematics and are constantly required to reflect on their reasoning. Definitions and theorems should be well motivated so that they are seen as helpful, powerful tools that make it easier to organize and understand mathematical ideas. Such classrooms will also benefit prospective teachers by serving as models for their own future classrooms. A corollary of this approach is that to emphasize mathematical reasoning, upper-division mathematics courses for teachers may need to spend more time on only a part of the traditional syllabus. The end of such a course could survey further material. Learning mathematical reasoning is more important than covering every possible topic.

Finally, learning mathematical reasoning and actively participating in class will be easier when the learning builds on existing knowledge of high school mathematics. For example, undergraduates have more experience to draw upon in an algebra class when discussing polynomial rings than non-commutative groups. Of course, building theories directly connected to high school mathematics can also strengthen and deepen prospective teachers' knowledge of what they will teach.

We begin with some suggestions for the courses in the short sequence outlined in the chapter appendix. There are two special suggestions that cut across all the rest—experiences that should be integrated across the entire spectrum of undergraduate mathematics:

> *Experience with reasoning and proof.* Reasoning is essential for all mathematical professions, especially for teaching. Making sense of mathematics makes it easier to understand, easier to teach, and intellectually satisfying for all students, including high school students who have no intention of going into technical fields. And proof is essential to all of mathematics. Accordingly, reasoning and proof, while it is often the focus of a particular

---

[5]For examples, see Wu, "Phoenix Rising," *American Educator*, 2011.

course, should be present, not always at the same level of rigor, in most undergraduate courses. For example, the Intermediate Value Theorem can be motivated and made plausible in a first calculus course; a more rigorous proof can be developed in a course in real analysis.

Experience with technology. Teachers should become familiar with various software programs and technology platforms, learning how to use them to analyze data, to reduce computational overhead, to build computational models of mathematical objects, and to perform mathematical experiments. The experiences should include dynamic geometry environments, computer algebra systems, and statistical software, used both to apply what students know and as tools to help them understand new mathematical ideas—in college, and in high school. Not only can the proper use of technology make complex ideas tractable, it can also help one understand subtle mathematical concepts. At the same time, technology used in a superficial way, without connection to mathematical reasoning, can take up precious course time without advancing learning.

COURSES TAKEN BY A VARIETY OF UNDERGRADUATE MAJORS

Single- and multi-variable calculus. The standard three-semester calculus sequence can help prospective teachers bring together many of the ideas in high school mathematics. They can derive results that may have been taken for granted in high school, such as the formulas for the volume of a cone and sphere, and they have an opportunity to master the ideas of algebra, clearing up common confusion among expressions, equations, and functions.

Calculus also opens up the arena of applied mathematics, deepening prospective teachers' understanding of mathematical modeling. Many calculus courses include a brief treatment of differential equations, providing prospective teachers an opportunity to see where a subject they teach in high school is heading in college.

Multi-variable calculus provides the same opportunity, opening up the subject into more serious applications to science and engineering than are available in single-variable calculus. Multi-variable calculus also provides essential background in analytic geometry. A careful treatment of geometry in $\mathbf{R}^2$ and $\mathbf{R}^3$ using dot product to extend notions of length and angle, and developing equations of lines and planes, is extremely useful background for high school teaching.

Introduction to linear algebra. After calculus, linear algebra is the most powerful, comprehensive theory that teachers will encounter. It is an excellent place to begin proving theorems because of the computational nature of many of its proofs, and provides an opportunity for teachers to experience the mathematical practice of abstracting a mathematical idea from many examples. A concrete course anchored in specific examples and contemporary applications is more likely to serve the needs of prospective teachers than a course in the theory of abstract vector spaces. Important examples include $\mathbf{R}^n$ and the vector space of polynomials on which differentiation and integration act as linear operators; contemporary applications such as regression, computer visualization, and web search engines.

The Common Core State Standards include operations on vectors and matrices and the use of matrices to solve systems of equations. Matrices and matrix algebra represent an important generalization of numbers and number algebra, providing an opportunity to reflect on the properties of operations as general rules for algebraic manipulation. The representation of complex numbers by matrices is a particularly relevant instance of this for high school teachers. In addition to this algebraic aspect of matrices, the geometric interpretation of matrices as transformations of the plane and three-space is also useful for prospective teachers, providing, for example, a connection between solving equations and finding inverse functions. Linear equations and functions are prominent in secondary school mathematics, and geometric interpretations of them in higher dimensions can deepen teachers' understanding of these notions. For example, the classification by dimension of solution sets of systems of linear equations gives perspective on the one-dimensional case, making the cases of no or infinitely many solutions to linear equations in one variable seem less exceptional.

Statistics and probability. The Common Core State Standards include interpretation of data, an informal treatment of inference, basic probability (including conditional probability), and, in the + standards,[6] the use of probability to make decisions. In preparation for teaching this, teachers should see real-world data sets, understand what makes a data set good or bad for answering the question at hand, appreciate the omnipresence of variability, and see the quantification and explanation of variability via statistical models that incorporate variability.

For this purpose, the standard statistics course that serves future engineers and science majors in many institutions may not be appropriate. A modern version, given as one course or a two-course sequence, centers around statistical concepts and real-world case studies, and makes use of technology in an active learning environment. It would contain the following topics: formulation of statistical questions; exploration and display of univariate data sets and comparisons among multiple univariate data sets; exploration and display of bivariate categorical data (two-way tables, association) and bivariate measurement data (scatter plots, association, simple linear regression, correlation); introduction to the use of randomization and simulation in data production and inferential reasoning; inference for means and proportions and differences of means or proportions, including notions of p-value and margin of error; and introduction to probability from a relative frequency perspective, including additive and multiplicative rules, conditional probability and independence. If given as a two-course sequence, it would include additionally the topics described on page 66.

COURSES INTENDED FOR ALL MATHEMATICS MAJORS

This section describes ways in which courses commonly occurring in the mathematics major can be geared to the needs of prospective teachers.

Introduction to proofs. In order to be able to recognize, foster, and correct their students' efforts at mathematical reasoning and proof, prospective high school teachers should analyze and construct proofs themselves, from simple derivations to proofs

---

[6]The CCSS standards for high school include standards marked with a +, indicating standards that are beyond the college- and career-ready threshold.

of major theorems. Also, they need to see how reasoning and proof occur in high school mathematics outside of their traditional home in axiomatic Euclidean geometry. Important examples include proof of the quadratic formula and derivation of the formula for the volume of a cone from an informal limiting argument that starts from the volume of a pyramid. Moreover, teachers must know that proof and deduction are used not only to convince but also to solve problems and gain insights. In particular, teachers need to see why solving equations is a matter of logical deduction and be able to describe the deductive nature of each step in solving an equation.

Prospective teachers can gain experience with reasoning and proof in a number of different courses, including a dedicated introduction to proofs course for mathematics majors, Linear Algebra, Abstract Algebra, Geometry, or a course on high school mathematics from an advanced standpoint. In the last course, polynomial algebra, geometry, number theory, and complex numbers are good venues for learning about proofs.

Abstract algebra. An advanced standpoint reveals much of high school mathematics as the algebra of rings and fields. Abstract algebra for prospective high school teachers should therefore emphasize rings and fields over groups. These structures underlie the base-ten arithmetic of integers and decimals, and operations with polynomials and rational functions. This course is an opportunity for prospective teachers to gain an understanding of how the properties of operations determine the permissible manipulations of algebraic expressions and to appreciate the distinction between these properties and "rules" that are merely conventions about notation (for example, the order of operations). Attention to the concepts of identity and inverse help prospective teachers see the concepts of multiplicative inverse, additive inverse, inverse matrix, and inverse function as examples of the same idea. Particularly important is the isomorphism between the additive group of the real numbers and the multiplicative group of the positive real numbers given by the exponential and logarithm functions, and the fact that the laws of exponents and the laws of logarithms are just two isomorphic collections of statements.

The study of rings also provides an opportunity for teachers to see that non-negative integers represented in base ten can be viewed as "polynomials in 10," and to consider ways in which polynomials might be considered as numbers in base x, as well as ways in which these analogies have shortcomings. Whole-number arithmetic can be viewed as a restriction of operations on polynomials to "polynomials in 10."

The division algorithm and the Euclidean algorithm for polynomials and integers, the Remainder Theorem, and the Factor Theorem are important for teachers, because these theorems underlie the algebra that they will teach.

It is also valuable for prospective teachers to see the historical development of methods for representing and performing numerical and symbolic calculations, and of formal structures for number systems.

Another example of a connection between abstract algebra and high school mathematics is the connection between $\mathbf{C}$ and $\mathbf{R}[x]$. It would be quite useful for prospective teachers to see how $\mathbf{C}$ can be "built" as a quotient of $\mathbf{R}[x]$ and, more generally, how splitting fields for polynomials can be gotten in this way. The quadratic formula, Cardano's method, and the algorithm for solving quartics by radicals can all be developed from a structural perspective as a preview to Galois theory, bringing some coherence to the bag of tricks for factoring and completing the square that are traditional in high school algebra. Indeed, this coherence is a major goal of the CCSS high school standards.

Many of the examples given in the description of the number theory course on page 61 can also serve as ingredients for an abstract algebra course geared to prospective high school teachers.

The short sequence in the chapter appendix contains an additional elective for all mathematics majors. Here are some suggestions for material that can be included in such courses. Much of this material is also suitable for the recommended three courses designed specifically for prospective teachers.

The real number system and real analysis. It is an often unstated assumption of high school mathematics that the real numbers exist and satisfy the same properties of operations as the rational numbers. Teachers need to know how to prove what is unstated in high school in order to avoid false simplifications and to be able to answer questions from students seeking further understanding. Thus, a construction of the real numbers, a proof that they satisfy the properties of operations (the CCSS term for the field axioms), and a proof that they satisfy the Completeness Axiom are necessary for teachers. A definition of continuity for a function of a real variable and a proof of the Intermediate Value Theorem provide the underpinnings of the graphical methods for solving equations that are taught to high school students. Thus, they are needed ingredients in teachers' backgrounds. A treatment of the real numbers can also include a treatment of their representation as infinite decimals, including an understanding of decimal expansions as an address system on the number line and an analysis of the periods of decimal expansions of rational numbers using modular arithmetic.

These topics provide opportunities to make use of original historical sources, which can motivate the theory and make it seem less disconnected from school mathematics.

Modeling. In many departments, there is a modeling course for mathematics majors. The Common Core State Standards include an emphasis on modeling in high school, and prospective teachers should have experience modeling rich real-world problems. This includes some aspects of quantitative literacy: the ability to construct and analyze statistical models; the ability to construct and analyze expressions, equations, and functions that serve a given purpose, derived either from a real-world context or from a mathematical problem, and to express them in different ways when the purpose changes; and the ability to understand the limitations of mathematical and statistical models and modify them when necessary.

Differential equations. It would be useful for calculus teachers to see where the subject is going. The traditional course for engineers, heavy on analytical techniques, is not the best choice for teachers. Rather they would benefit from a course including quantitative and qualitative methods, and experience constructing and interpreting classical differential equations arising in science, possibly including partial differential equations. This is also an excellent course in which to include some historical material.

Group theory. A number of applications of group theory are useful for teachers. One is the application of groups to understand the permutations of the roots of an equation that leave the coefficients fixed.[7] A concrete introduction to Galois theory helps place the quadratic formula into a larger theoretical perspective. A second application is connected to the CCSS approach to geometry through rigid motions and dilations. Study of groups and group structures of transformations, and even the isometries of polygons and polyhedra, is very useful background for high school teachers as they integrate this transformational approach into their geometry teaching. A third application is the affine group of the line, that is, the group of transformations of the form $f(x) = ax + b$ for real numbers a and b, which underlies much of the work in high school algebra on transformations of graphs of functions, among other things.[8]

Number theory. Modular arithmetic lends itself to extensions of numerous topics appropriate for STEM-intending high school students, from the analysis of periods of decimal expansions of rational numbers to an understanding of public key cryptography.

A comparison of arithmetic in $\mathbf{Z}$ and $\mathbf{Z}/n\mathbf{Z}$ helps teachers understand the importance of the lack of zero divisors when teaching the "factor to solve" techniques for quadratic and higher-degree equations. It also provides examples of two different polynomials that define the same function, reinforcing the distinction between polynomials and polynomial functions.

A detailed examination of the parallel between $\mathbf{Z}$ and $\mathbf{Q}[x]$, showing, for example, the close connection between the Chinese Remainder Theorem and Lagrange interpolation, would help teachers tie together two of the main algebraic structures in pre-college mathematics.

Although high school mathematics curricula often mention irrational numbers, there are not many cases where numbers are proven irrational. The proof of unique factorization often included in a number theory course allows a proof that any fractional power of any rational number is irrational, unless it is obviously rational.

Many of the examples given in the description of the abstract algebra course can also serve as ingredients for a number theory course geared to prospective high school teachers.

History of mathematics. The history of mathematics can either be woven into existing mathematics courses or be presented in a mathematics course of its own. In both instances, it is important that the history be accurate; instructors who have no contact with historians need to be aware that findings from historical research may contradict popular accounts.

The history of mathematics can be used to raise some general issues about mathematics, such as the role of axiomatic systems, the nature of proof, and perhaps most importantly, mathematics as a living and evolving subject.[9]

History can illustrate the significance of notation. In medieval Europe, computations were made with counters or an abacus, and recorded with Roman numerals. By the sixteenth century, arithmetic was frequently done using Hindu-Arabic

---

[7]From a modern viewpoint, this is an application, but the notion of group arose in this context. See Grattan-Guinness's discussion of "irresolving the quintic" in *The Rainbow of Mathematics: A History of the Mathematical Sciences*, Norton, 1997.

[8]See, e.g., Howe, "The Secret Life of the $ax + b$ Group" in the web resources.

[9]These and other ideas are listed in Kleiner's "The Teaching of Abstract Algebra: An Historical Perspective" in *Learn From the Masters!*, MAA, 1995.

numerals—suggesting base-ten notation's affordances for written computation.[10] An innovation that occurred just before 1600 was the representation of quantities by using letters systematically rather than as abbreviations.[11] The power of the symbolic algebra that developed is suggested by the enormous mathematical and scientific progress of the following century, and the cumbersome nature of the preceding "rhetorical algebra."[12]

It is particularly useful for prospective high school teachers to work with primary sources. Working with primary sources gives practice in listening to "wrong" ideas. Primary documents show how hard some ideas have been, for example, the difficulties that Victorian mathematicians had with negative and complex numbers helps prospective teachers appreciate how hard these ideas can be for students who encounter them for the first time. Finally, primary documents exhibit older techniques, and so give an appreciation of how mathematics was done and how mathematical ideas could have developed.

The array of undergraduate electives in mathematics around the country is vast; the list above gives some examples, showing how existing electives can be used to help prepare majors for the profession of high school teaching.

COURSES DESIGNED PRIMARILY FOR PROSPECTIVE TEACHERS

As part of the mathematics major for prospective teachers, Recommendation 1 in Chapter 3 calls for three courses with a primary focus on high school mathematics from an advanced viewpoint.

Many of the electives described in the preceding section can, with some modification, meet the goals of this recommendation. Indeed, at many universities, enrollments in teacher preparation programs are too low to create special courses (and many majors don't decide to become teachers until late in their undergraduate program). Dual purpose electives can meet this recommendation if they meet the dual criterion of developing content expertise and reasoning skills described at the beginning of this chapter.

Instructors who design and teach these courses find them mathematically satisfying. Arnold Ross held up "thinking deeply about simple things" as a beacon in our discipline that shines brightly when we consider the mathematics of high school—bringing out its essential coherence and fitting it into the larger landscape of mathematics.

Specialized courses for prospective teachers need not follow a particular theme or format. Practicing high school teachers with insight about the mathematics that is useful to them in their profession can help in the design. Organizing principles that have been useful are:

---

[10]See Smith & Karpinski, *Hindu-Arabic Numerals*, Ginn and Company, 1911, pp. 136–137.

[11]This distinction is illustrated by x^2 and x vs. sq and rt (or square and root).

[12]For details of previous and subsequent notations, see Cajori, *A History of Mathematical Notations*, Dover, 1993. A similarity between base-ten notation and symbolic algebra is that they are "action notations" in which computations can occur, rather than "display notations" that only record results. See Kaput, "Democratizing Access to Calculus," *Mathematical Thinking and Problem Solving*, Erlbaum, 1994, p. 101.

Treat high school mathematics from an advanced standpoint. Courses following this principle should emphasize the inherent coherence of the mathematics of high school, the structure of mathematical ideas from which the high school syllabus is derived.

Take up a particular mathematical terrain related to high school mathematics and develop it in depth. For example, a course might develop the mathematics necessary to prove the fundamental theorem of algebra or the impossibility of the classical straight-edge and compass constructions.

Develop mathematics that is useful in teachers' professional lives. Courses following this principle might take up classical ideas that are not normally included in a mathematics major but are of special use for teachers, such as the classical theory of equations or three-dimensional Euclidean geometry.

Here are some examples of courses and course sequences that illustrate these principles:

A treatment of the Pythagorean Theorem and Pythagorean triples that leads to the unit circle, trigonometry and Euclidean formulas for areas of triangles, leading in turn to polygonal approximations of the unit circle and π (one-semester course).[13]

The theory of rational numbers based on the number line, the Euclidean algorithm, complex numbers and the Fundamental Theorem of Algebra, roots and factorizations of polynomials, Euclidean geometry (including congruence and similarity), geometric transformations, axiomatic systems, basic trigonometry, equation, functions, graphs (three-semester sequence).[14]

Classical number systems, starting with the natural numbers and progressing through the integral, rational, real, and complex number systems; mathematical systems, including operations within fields and the foundations of the real number system; modern Euclidean and non-Euclidean geometry, including topics in plane and solid geometry, the axioms of Euclidean, projective, and non-Euclidean geometry; and an introduction to group theory: permutation groups, cyclic groups, theory of finite groups, group homomorphisms and isomorphisms, and abelian groups (five-quarter sequence).[15]

Here are some ideas for the ingredients of specialized courses for teachers:

Geometry and transformations. The approach to geometry in the Common Core State Standards replaces the initial phases of axiomatic Euclidean geometry. In the latter, the triangle congruence and similarity criteria are derived from axioms. The Common Core, on the other hand, uses a treatment based on translations, rotations, reflections, and dilations, whose basic angle and distance preserving properties are taken as axiomatic. It is essential that teachers see a detailed exposition of this development.

---

[13]Rotman's *Journey Through Mathematics* has been used for such a course.

[14]This description is based on the University of California at Berkeley courses 151, 152, 153.

[15]This description is based on the University of California, Santa Barbara courses 101A-B, 102A-B, 103.

The Pythagorean Theorem is a fundamental topic in school geometry, and students should see a proof of the theorem and its converse. It can also provide an excursion into some number theory topics such as generating Pythagorean triples, the Congruent Number Problem, and Fermat's Last Theorem, the last two being examples of how questions arising in high school can lead to the frontiers of current research.

An understanding of the role played by the parallel postulate in Euclidean geometry is essential for geometry teachers. Knowing where the postulate is hiding underneath the major theorems in plane geometry, from angle sums in polygons to area formulas, helps teachers build a coherent and logical story for their students. It can also be helpful to know that there are geometries in which the parallel postulate does not hold.

There are many classical results in geometry that may fall through the cracks in undergraduate preparation. Everything from Heron's (and Brahmagupta's) formula to Ptolemy's theorem to the theory of cyclic quadrilaterals can, in the hands of a well-prepared teacher, enrich high school geometry.

Analytic geometry. Many connections between high school topics and the content of undergraduate mathematics can be highlighted in a course in analytic geometry. For example, teachers would benefit from analytic and vector proofs of standard geometric theorems from high school (showing that a line in the coordinate plane is characterized by constant slope is highlighted in the Common Core State Standards[16]). The complete analysis of the graph of a quadratic equation in two variables gives teachers tools that they can use in their high school classes, and it is also a concrete example of the use of the theory of eigenvalues and eigenvectors that they may have studied in linear algebra. Applications of conic sections to everything from quadratic forms to optics shows the power of some of these classical methods.

Complex numbers and trigonometry. Complex numbers can fall into the chasm between high school and college, with high school teachers assuming they will be taught in college and college instructors assuming they have been taught in high school. And the trigonometry that students learn in high school often ends up being a jumble of facts and techniques, with little texture and coherence. Treating complex numbers and trigonometry together can prepare teachers with a solid foundation in both areas, and it can help them make sense of some seemingly disconnected terrains in upper-level high school mathematics.

An understanding of the historical evolution of the complex number system—that complex numbers evolved from a mysterious tool used to solve cubic equations with real coefficients and roots to one of the most useful structures in mathematics—is extremely valuable for high school teachers. For example, $\mathbf{C}$ allows one to connect algebra and geometry in ways that are very hard to see otherwise; de Moivre's Theorem establishes a tight connection between the algebraic structure of roots of the equation and the geometry of the regular n-gon. And there's an equally tight connection between complex numbers and trigonometry—many trigonometric identities can be viewed as algebraic identities in $\mathbf{C}$.

---

[16]Part of an eighth grade standard is: "Use similar triangles to explain why the slope m is the same between any two distinct points on a non-vertical line in the coordinate plane."

Teachers would also benefit from seeing applications of complex numbers, ranging from how they underlie electromagnetic communication such as cell phones to how they can be used to generate Pythagorean triples.

In another direction, prospective teachers could bring some needed parsimony and coherence to the high school trigonometry that they will teach if they understand that a small number of basic ideas—the invariance of the unit circle under rotation and the Pythagorean Theorem, for example—can be used to generate all of the results in elementary trigonometry, from the formulas for the sine and cosine of sums and differences to the relationships among co-functions.

Research experience. In all mathematical professions, experts are able to view the discipline from several perspectives. Expert high school teachers should know mathematics in at least four overlapping ways:

> As *scholars.* They should have a solid grounding in classical mathematics, including major results and applications, the history of ideas, and connections to pre-college mathematics.

> As *educators.* They should understand the habits of mind that underlie different branches—arithmetic, algebra, geometry, analysis, modeling, and statistics—and how these develop in learners.

> As *mathematicians.* They should have experienced a sustained immersion in mathematics that includes performing experiments and grappling with problems, building abstractions as a result of reflection on the experiments, and developing theories that bring coherence to the abstractions.

> As *teachers.* They should be expert in uses of mathematics that are specific to the profession, e.g., finding simple ways to make mathematics tractable for beginners; the craft of task design, the ability to see underlying themes and connections in school mathematics, and the mining of student ideas.

Much of the education that prospective high school teachers get as undergraduates focuses on knowing mathematics as a scholar (in mathematics departments) and knowing mathematics as an educator (in education departments). Few undergraduates get a chance to develop knowledge of mathematics as a mathematician or as a teacher. Knowing mathematics as a mathematician is important for prospective high school teachers (in fact, for any mathematics major). It is possible that the discontinuity between how mathematicians and teachers view the whole enterprise of mathematics—what is important, what is convention, what constitutes expertise, and even what it means to understand the subject—is because the typical mathematics major does not provide an intense immersion experience in mathematics. Teachers who have engaged in a research-like experience for a sustained period of time frequently report that it greatly affects what they teach, how they teach, what they deem important, and even their ability to make sense of standard mathematics courses.[17] The research experiences available in many departments and summer

---

[17]For example, see the reports of Focus on Mathematics (a Math Science Partnership). Comments from teachers include: "Study groups have made 'asking the next question' a much

programs are recommended for prospective mathematics teachers. (See the web resources for examples.)

Important Additional Mathematics

It is impossible to learn all the mathematics one will use in any mathematical profession, including teaching, in four years of college. Therefore teachers will need opportunities to learn further topics throughout their careers. This section describes important additional mathematics that can be the content of undergraduate electives, graduate courses for prospective and practicing teachers, or professional development programs for practicing teachers.

The involvement of the mathematical community in career-long, content-based professional development programs for practicing teachers provides an opportunity for mathematicians and statisticians to have a profound effect on the content and direction of high school mathematics. And it provides teachers with years of opportunities to learn more mathematics and statistics that is especially useful in their profession and to be partners with mathematicians and statisticians in a desperately needed effort to improve professional development experiences. Such in-service should be offered at times of the day and year that allow teachers to participate, such as during the summer, evenings, or weekends.

All of the topics listed in the previous section as possible ingredients for specialized courses for prospective teachers are also fair game for constructing in-service courses. Here are some additional ideas:

Further statistics. For teachers who plan to teach statistics, including high school courses that address the more advanced parts of the statistics standards in the CCSS or AP courses, a second course is recommended. Suggested topics include: regression analysis, including exponential and quadratic models; transformations of data (logs, powers); categorical data analysis, including logistic regression and chi-square tests; introduction to study design (surveys, experiments, and observational studies); randomization procedures for data production and inference; and introduction to one-way analysis of variance.

Discrete mathematics and computer science. Many states are beginning to require a fourth year of high school mathematics. Not all students will be inclined or able to satisfy this requirement with precalculus, calculus, or statistics. States are developing additional courses that build on the modeling and + standards in the CCSS. Teachers of modeling courses will benefit from courses that include topics such as the basics of graph theory; finite difference equations, iteration and recursion; the Binomial Theorem and its use in algebra and probability; and computer programming.

Further geometry. Geometry teachers could profitably study geometric limit problems of the sort studied in ancient Greece, for example the method used to determine the area of a disk. Other possible topics include geometric optimization (finding shortest paths, for example); equi-decomposibility, area, and volume; non-Euclidean geometries; axiomatic approaches to geometry; and a brief introduction

---

more intriguing mathematical exploration than I previously had imagined or realized I could access," *Focus on Mathematics Summative Evaluation Report 2009*, p. 29.

to computational geometry. The latter can draw heavily on high school geometry; for example, questions such as: Given the three-dimensional coordinates of an observer and the corners of a tetrahedron, which faces of the tetrahedron can the observer see?

Further algebra. There are applications of abstract algebra that are especially useful for high school teachers. These include straight-edge and compass constructions, solvability of equations by radicals, and applications of cyclotomy and roots of unity in geometry. Rational points on conics and norms from quadratic fields can be applied to the problem of creating problems for students that have integer solutions. An introduction to algebraic geometry can help teachers bring some coherence to the analytic geometry they teach in precalculus, helping them make deeper connections between the algebra of polynomials and the graphs of polynomial curves.

Further history of mathematics. Many topics in the history of mathematics are closely related to high school mathematics, for example, history of statistics, history of trigonometry, and history of (premodern) algebra. It is important to make sure that the materials used for courses on these topics include a significant amount of mathematical content.

Further study of the mathematics of high school. Teachers should study the mathematics of high school in their undergraduate programs, as suggested above. Further coursework that focuses on the mathematics they are teaching and how it fits into the broader landscape of mathematics is valuable. Especially important in such coursework is a goal of bringing mathematical coherence to high school mathematics, showing how a few general-purpose ideas and methods can be used across the entire high school spectrum of topics, replacing much of the special-purpose paraphernalia that clutters many high school programs, such as the various mnemonics for formulas in trigonometry. Such courses can fit in graduate degree programs for teachers offered by mathematics departments alone or in conjunction with education departments.

Other advanced topics. The terrain is vast—much mathematics and statistics is missed in undergraduate programs simply because of lack of time, and much of this can bring new insights into high school topics. Applications of the arithmetic-geometric mean inequality to optimization problems, the use of measure theory to connect area and probability, the irrationality or transcendence of the classical constants from algebra and geometry, the famous impossibility theorems (squaring the circle, trisecting the angle), Gödel's Incompleteness Theorems, properties of iterated geometric constructions, Hilbert's axioms for area and volume, and so many other areas can help teachers learn mathematics that is useful in their work and at the same time important in the field.

Essential Experiences for Practicing Teachers

All teachers need continuing opportunities to deepen and strengthen their mathematical knowledge for teaching. Many teachers prepared before the era of the CCSS will need opportunities to study content that they have not previously taught, particularly in the areas of statistics and probability.

In addition to learning more mathematical topics, teachers need experiences that renew and strengthen their interest in and love for mathematics, help them represent mathematics as a living discipline to their students by exemplifying mathematical practices, figure out how to pose tasks to students that highlight the essential ideas under consideration, to listen to and understand students' ideas, and to respond to those ideas and point out flaws in students' arguments. Being able to place themselves in the position of mathematics learners can help them think about their students' perspectives. These needs create opportunities for mathematics departments to participate in the creation of important professional development experiences for high school mathematics teachers.

The research experiences described above for prospective teachers can also be important for practicing teachers. Here are some additional ideas:

Math teachers' circles and study groups. Math teachers' circles, in which teachers and mathematicians work together on interesting mathematics, provide ongoing opportunities for teachers to develop their mathematical habits of mind while deepening their understanding of mathematical connections and their appreciation of mathematics as a creative, open subject. Unlike more structured courses, math teachers' circles are informal sessions that meet regularly and can include the same participants for multiple years. A substantial benefit of such programs is that they address the isolation of both high school teachers and practicing mathematicians: they establish communities of mathematical practice in which teachers and mathematicians can learn about each others' profession, culture, and work.

Immersion experiences. For all the reasons discussed earlier under "research experience" (p. 65), teaching mathematics is greatly enhanced when teachers work themselves as mathematicians and statisticians. For practicing teachers, an immersion experience (usually over a summer) in which one works on a small, low-threshold, high-ceiling cluster of ideas for a sustained period of time has profession-specific benefits. For example, it helps teachers understand the nature of doing mathematics and statistics, it reminds them that frustration, confusion, and struggle are all natural parts of being a learner, it helps them connect ideas that seem on the surface to be quite different, it shows the value of refining ill-formed ideas through the use of precise language, and it keeps alive the passion for mathematics that was ignited in undergraduate school. Mathematicians and statisticians are the ideal resources to help design and implement such immersion programs.

Lesson study. In lesson study, teachers work in small teams including fellow teachers, mathematicians, mathematics educators, and administrators. The teams carefully and collaboratively craft lesson plans designed to meet both content goals and general learning or affective goals for students—such as working together to solve problems or being excited to learn about nature. One or more members of the team teaches the lesson while the other team members observe the lesson implementation.

The team then debriefs the lesson and makes revisions, sometimes teaching the revised lesson to another group of students. Mathematicians can play an important role as part of a lesson study team in helping to think flexibly about the mathematical goals of the lesson, tasks to include, mathematical issues to address following observation of the lesson in action, and mathematical issues to consider when revising the lesson. Working as part of a team, mathematicians, along with others, can bring expertise to bear on the interdisciplinary work of teaching. Hosting a high school lesson study group on teaching topics in trigonometry or precalculus can have an added benefit for instructors of related courses at universities or two- and four-year colleges.

Chapter 6 Appendix: Sample Undergraduate Mathematics Sequences

Short sequence (33 semester-hours).

I Courses taken by undergraduates in a variety of majors (15+ semester-hours)
   – Single- and Multi-variable Calculus (9+ semester-hours)
   – Introduction to Linear Algebra (3 semester-hours)
   – Introduction to Statistics (3 semester-hours)

II Courses intended for all mathematics majors (9 semester-hours)
   – Introduction to Proofs (3 semester-hours)
   – Abstract Algebra (approach emphasizing rings and polynomials) (3 semester-hours)
   – A third course for all mathematics majors (e.g., Differential Equations) (3 semester-hours)

III Courses designed primarily for prospective teachers (9 semester-hours).

Long sequence (42 semester-hours).

I Courses taken by undergraduates in a variety of majors (21 semester-hours)
   – Single- and Multi-variable Calculus (9+ semester-hours)
   – Introduction to Linear Algebra (3 semester-hours)
   – Introduction to Computer Programming (3 semester-hours)
   – Introduction to Statistics I, II (6 semester-hours)

II Courses intended for all mathematics majors (12 semester-hours)
   – Introduction to Proofs (3 semester-hours)
   – Advanced Calculus (3 semester-hours)
   – Abstract Algebra (approach emphasizing rings and polynomials) (3 semester-hours)
   – Geometry or Mathematical Modeling (3 semester-hours)

III Courses designed primarily for prospective teachers (9 semester-hours).

Selected References and Information Sources

This annotated list describes recent reports that inform MET II's recommendations, and gives sources of information about accreditation and licensure.

Early Childhood: Teacher Preparation and Professional Development

Mathematics Learning in Early Childhood: Paths Toward Excellence and Equity, National Research Council, 2009, `http://www.nap.edu/catalog.php?record_id=12519`

This report summarizes research concerned with early childhood teaching and learning of mathematics. It notes that:

> Traditionally, early childhood educators have been taught that mathematics is a subject that requires the use of instructional practices that are developmentally inappropriate for young children. (p. 299)

> The content of young children's mathematics can be both deep and broad, and, when provided with engaging and developmentally appropriate mathematics activities, their mathematics knowledge flourishes. Yet these research findings are largely not represented in practice. (p. 300)

Much research on teaching–learning paths focuses on early childhood, and its findings are described in this report. These have implications for several aspects of early childhood education and the report gives recommendations for curriculum, instruction, and standards. The recommendations about preparation and professional development for the early childhood workforce are especially relevant to MET II. These are:

> Coursework and practicum requirements for early childhood educators should be changed to reflect *an increased emphasis on children's mathematics* as described in the report. These changes should also be made and enforced by early childhood organizations that oversee credentialing, accreditation, and recognition of teacher professional development programs. (pp. 3–4, emphasis added)

> An essential component of a coordinated national early childhood mathematics initiative is the provision of professional development to early childhood in-service teachers that helps them

(a) *to understand the necessary mathematics*, the crucial teaching–learning paths, and the principles of intentional teaching and curriculum and (b) to learn how to implement a curriculum. (p. 3, emphasis added)

Elementary Mathematics Specialists: Preparation and Certification

Standards for Elementary Mathematics Specialists: A Reference for Teacher Credentialing and Degree Programs, Association for Mathematics Teacher Educators, 2010, `http://www.amte.net/resources/amte-documents`

This report notes: "Many have made the case that practicing elementary school teachers are not adequately prepared to meet the demands for increasing student achievement in mathematics." Elementary mathematics specialists are an "alternative to increasing all elementary teachers' content knowledge (a problem of huge scale) by focusing the need for expertise on fewer teachers."

Depending on location, an elementary mathematics specialist may have the title elementary mathematics coach, elementary mathematics instructional leader, mathematics support teacher, mathematics resource teacher, mentor teacher, or lead teacher. In several states, specialists and mathematicians collaborate in teaching courses offered for teachers in the specialists' districts.

This report summarizes research on specialists' effectiveness and outlines the knowledge, skills, and leadership qualities necessary for their roles and responsibilities. It is intended as a starting point for state agencies in establishment of certification guidelines and as a guide for institutions of higher education in creation of programs to prepare specialists.

Teacher Preparation

Preparing Teachers: Building Sound Evidence for Sound Policy, National Research Council, 2010, `http://www.nap.edu/catalog/12882.html`

This report summarizes what is known about teacher preparation, in general and with respect to teaching mathematics, concluding that:

> Current research and professional consensus correspond in suggesting that all mathematics teachers ... rely on: mathematical knowledge for teaching, that is, knowledge not just of the content they are responsible for teaching, but also of the broader mathematical context for that knowledge and the connections between the material they teach and other important mathematics content. (pp. 114–115)

Postsecondary institutions predominate in preparing teachers, educating 70% to 80% of those who complete a preparation program. There are numerous alternative pathways for teacher preparation. These include "fellows' programs" established by school districts, which usually combine expedited entrance into teaching with tuition-supported enrollment in graduate study in education.

Information about what these programs do is sparse, however, the report concludes that "there is relatively good evidence that mathematics preparation for prospective teachers provides insufficient coursework in mathematics as a discipline and mathematical pedagogy" (p. 123).

Moreover, the mathematics that teachers need to know is in sharp contrast with state requirements for licensure.

> 33 of the 50 states and the District of Columbia require that high school teachers have majored in the subject they plan to teach in order to be certified, but only 3 states have that requirement for middle school teachers (data from 2006 and 2008; see `http://www.edcounts.org` [February 2010]). Forty-two states require prospective teachers to pass a written test in the subject in which they want to be certified, and six require passage of a written test in subject-specific pedagogy.

> Limited information is available on the content of teacher certification tests. A study of certification and licensure examinations in mathematics by the Education Trust (1999) reviewed the level of mathematics knowledge necessary to succeed on the tests required of secondary mathematics teachers. The authors found that the tests rarely assessed content that exceeded knowledge that an 11th or 12th grader would be expected to have and did not reflect the deep knowledge of the subject one would expect of a college-educated mathematics major or someone who had done advanced study of school mathematics. Moreover, the Education Trust found that the cut scores (for passing or failing) for most state licensure examinations are so low that prospective teachers do not even need to have a working knowledge of high school mathematics in order to pass. Although this study is modest, its results align with the general perception that state tests for teacher certification do not reflect ambitious conceptions of content knowledge. (p. 118)

Professional Development

Key State Education Policies on PK–12 Education: 2008, Council of Chief State School Officers, `http://www.ccsso.org/Resources/Publications.html`

An overview of individual state policies on professional development is given on pp. 22–24. Professional development requirements are specified by 50 states. The majority require 6 semester-hours of professional development over approximately 5 years. Twenty-four of these states specify that professional development should be aligned with state content standards.

Effects of Teacher Professional Development on Gains in Student Achievement, Council of Chief State School Officers, 2009, `http://www.ccsso.org/Resources/Publications.html`

Few studies of professional development use an experimental or quasi-experimental research design. This report gives a systematic analysis of 16 studies that did. Two of these covered the Northeast Front Range Math Science Partnership (whose focus was science). Twelve studies focused on mathematics. Common patterns of successful professional development programs are summarized on p. 27:

- strong emphasis on teachers learning specific subject content as well as pedagogical content for how to teach the content to students.

- multiple activities to provide follow-up reinforcement of learning, assistance with implementation, and support for teachers from mentors and colleagues in their schools.

- duration: 14 of the 16 programs continued for six months or more. The mean contact time with teachers in program activities was 91 hours.

Designing for Sustainability: Lessons Learned About Deepening Teacher Content Knowledge from Four Cases in NSF's Math and Science Partnership Program, Horizon Research, 2010, `http://www.mspkmd.net/cases/tck/sustainability/crosscase.pdf`

This report elaborates and illustrates lessons learned from experiences of the Math Science Partnerships. Page 8 lists these as:

- Recognize that it takes time to develop and nurture a *productive partnership.*

- Consider how to *engage a range of important stakeholders whose support is important* for efforts to deepen teacher content knowledge.

- Help ensure that *key policies in the system are aligned with the vision* underlying the reform efforts.

- Design and implement professional development that is not only aligned with the project goals, but is also both feasible and *likely to be effective with the teachers in their particular context.*

- *Use data* to inform decisions, improve the quality of the interventions, and provide evidence to encourage support for system change.

- Work to *develop capacity and infrastructure* to strengthen teachers' content knowledge and pedagogical skills, both during the funded period and beyond.

National Impact Report: Math and Science Partnership Program, National Science Foundation, 2010, `http://hub.mspnet.org/index.cfm/20607`

This report gives an overview of the National Science Foundation's Math Science Partnership program and its impact. Some features that may be of particular interest to MET II readers are:

- Yearly score increases between 2004 and 2009 on the 11th grade mathematics exam of the Texas Assessment of Knowledge and Skills for students of teachers who participated in an MSP mathematics leadership institute (p. 6).

- Yearly score increases between 2003 and 2007 on state assessments for students in schools that participated in MSP projects (pp. 10–11).

- Five-year score increases for elementary students in schools that were significantly involved in MSP projects (p. 12).

- Discussion of changes in university policies to reduce barriers to faculty involvement in activities for increasing K–12 student achievement (p. 15).

Supporting Implementation of the Common Core State Standards for Mathematics: Recommendations for Professional Development, Friday Institute for Educational Innovation at the North Carolina State University College of Education, 2012, `http://www.amte.net/resources/ccssm`

These recommendations are intended to support large-scale, system-level implementation of professional development (PD) initiatives aligned with the CCSS. These rest on four principles of effective PD derived from research listed on p. 7 of the report:

- PD should be intensive, ongoing, and connected to [teaching] practice.

- PD should focus on student learning and address the teaching of specific content.

- PD should align with school improvement priorities and goals.

- PD should build strong working relationships among teachers.

Credentials and Accreditation

Significantly different new accreditation standards for preparation programs are forthcoming from the *Council for the Accreditation of Educator Preparation*. This organization was formed by the merger of the National Council for the Accreditation of Teacher Education (NCATE) and the Teacher Education Accreditation Council (TEAC), `http://www.caepsite.org`

The Council for Exceptional Children gives information about program accreditation and licensure for special education teachers, `http://www.cec.sped.org`

The Association for Middle Level Education lists middle level teacher certification/licensure patterns by state, `http://www.amle.org`

The Elementary Mathematics Specialists and Teacher Leaders Project lists mathematics specialist certifications and endorsements by state, `http://mathspecialists.org`

The National Board for Professional Teaching Standards offers an advanced teaching credential in 25 different areas, `http://www.nbpts.org`

These credentials complement, but do not replace, a state's teacher license. The certificate areas that include mathematics are:

- Early childhood (ages 3–8)

- Middle childhood (ages 7–12)

- Mathematics (ages 11–18+)

- Exceptional needs (ages birth to 21+)

The Common Core State Standards: Overview of Content

Grades K–8

At each K–8 grade, the standards are grouped in clusters. The clusters are organized in mathematical domains, which may span multiple grade levels. This appendix lists the names of these domains and clusters.

Note that the same cluster name may occur at different grades. However, this does not indicate that the standards in those clusters are identical at different grade levels. For example, the cluster "Add and subtract within 20" at grade 1 includes "Add and subtract within 20, demonstrating fluency for addition and subtraction within 10." At grade 2, the same cluster name includes "Fluently add and subtract within 20, using mental strategies."

Grade	Counting and Cardinality	Operations and Algebraic Thinking	Number and Operations in Base Ten	Measurement and Data	Geometry
K	• Know number names and the count sequence. • Count to tell the number of objects. • Compare numbers.	• Understand addition as putting together and adding to, and understand subtraction as taking apart and taking from.		• Describe and compare measurable attributes. • Classify objects and count the number of objects in categories.	• Identify and describe shapes. • Analyze, compare, create, and compose shapes.
1		• Represent and solve problems involving addition and subtraction. • Understand and apply properties of operations and the relationship between addition and subtraction. • Add and subtract within 20. • Work with addition and subtraction equations.	• Extend the counting sequence. • Understand place value. • Use place value understanding and properties of operations to add and subtract.	• Measure lengths indirectly and by iterating length units. • Tell and write time. • Represent and interpret data.	• Reason with shapes and their attributes.
2		• Represent and solve problems involving addition and subtraction. • Add and subtract within 20. • Work with equal groups of objects to gain foundations for multiplication.	• Work with numbers 11–19 to gain foundations for place value.	• Measure and estimate lengths in standard units. • Relate addition and subtraction to length. • Work with time and money. • Represent and interpret data.	• Reason with shapes and their attributes.

	Number and Operations—Fractions	Operations and Algebraic Thinking	Number and Operations in Base Ten	Measurement and Data	Geometry
3	• Develop understanding of fractions as numbers.	• Represent and solve problems involving multiplication and division. • Understand properties of multiplication and the relationship between multiplication and division. • Multiply and divide within 100. • Solve problems involving the four operations, and identify and explain patterns in arithmetic.	• Use place value understanding and properties of operations to perform multi-digit arithmetic.	• Solve problems involving measurement and estimation of intervals of time, liquid volumes, and masses of objects. • Represent and interpret data. • Geometric measurement: understand concepts of area and relate area to multiplication and to addition. • Geometric measurement: recognize perimeter as an attribute of plane figures and distinguish between linear and area measures.	• Reason with shapes and their attributes.
4	• Extend understanding of fraction equivalence and ordering. • Build fractions from unit fractions by applying and extending previous understandings of operations on whole numbers. • Understand decimal notation for fractions, and compare decimal fractions.	• Use the four operations with whole numbers to solve problems. • Gain familiarity with factors and multiples. • Generate and analyze patterns.	• Generalize place value understanding for multi-digit whole numbers. • Use place value understanding and properties of operations to perform multi-digit arithmetic.	• Solve problems involving measurement and conversion of measurements from a larger unit to a smaller unit. • Represent and interpret data. • Geometric measurement: understand concepts of angle and measure angles.	• Draw and identify lines and angles, and classify shapes by properties of their lines and angles.
5	• Use equivalent fractions as a strategy to add and subtract fractions. • Apply and extend previous understandings of multiplication and division to multiply and divide fractions.	• Write and interpret numerical expressions. • Analyze patterns and relationships.	• Understand the place value system. • Perform operations with multi-digit whole numbers and with decimals to hundredths.	• Convert like measurement units within a given measurement system. • Represent and interpret data. • Geometric measurement: understand concepts of volume and relate volume to multiplication and to addition.	• Graph points on the coordinate plane to solve real-world and mathematical problems. • Classify two-dimensional figures into categories based on their properties.

	Ratios and Proportional Relationships	Expressions and Equations	The Number System	Statistics and Probability	Geometry
6	• Understand ratio concepts and use ratio reasoning to solve problems.	• Apply and extend previous understandings of arithmetic to algebraic expressions. • Reason about and solve one-variable equations and inequalities. • Represent and analyze quantitative relationships between dependent and independent variables.	• Apply and extend previous understandings of multiplication and division to divide fractions by fractions. • Compute fluently with multi-digit numbers and find common factors and multiples. • Apply and extend previous understandings of numbers to the system of rational numbers.	• Develop understanding of statistical variability. • Summarize and describe distributions.	• Solve real-world and mathematical problems involving area, surface area, and volume.
7	• Analyze proportional relationships and use them to solve real-world and mathematical problems.	• Use properties of operations to generate equivalent expressions. • Solve real-life and mathematical problems using numerical and algebraic expressions and equations.	• Apply and extend previous understandings of operations with fractions to add, subtract, multiply, and divide rational numbers.	• Use random sampling to draw inferences about a population. • Draw informal comparative inferences about two populations. • Investigate chance processes and develop, use, and evaluate probability models.	• Draw, construct and describe geometrical figures and describe the relationships between them. • Solve real-life and mathematical problems involving angle measure, area, surface area, and volume.
8	**Functions** • Define, evaluate, and compare functions. • Use functions to model relationships between quantities.	• Work with radicals and integer exponents. • Understand the connections between proportional relationships, lines, and linear equations. • Analyze and solve linear equations and pairs of simultaneous linear equations.	• Know that there are numbers that are not rational, and approximate them by rational numbers.	• Investigate patterns of association in bivariate data.	• Understand congruence and similarity using physical models, transparencies, or geometry software. • Understand and apply the Pythagorean Theorem. • Solve real-world and mathematical problems involving volume of cylinders, cones and spheres.

High School

The standards for high school are organized in conceptual categories, which may span multiple courses. Within these categories, related standards are grouped together. The categories and group headings are listed below. In the CCSS, additional mathematics that STEM-intended students should learn is indicated by (+) on individual standards. In the listing below, group headings for which all standards of this type are indicated by (+). Similarly, group headings for which all standards are modeling standards are indicated by (⋆).

Number and Quantity

The Real Number System

- Extend the properties of exponents to rational exponents
- Use properties of rational and irrational numbers

Quantities⋆

- Reason quantitatively and use units to solve problems

The Complex Number System

- Perform arithmetic operations with complex numbers
- Represent complex numbers and their operations on the complex plane +
- Use complex numbers in polynomial identities and equations

Vector and Matrix Quantities +

- Represent and model with vector quantities
- Perform operations on vectors
- Perform operations on matrices and use matrices in applications

Algebra

Seeing Structure in Expressions

- Interpret the structure of expressions
- Write expressions in equivalent forms to solve problems

Arithmetic with Polynomials and Rational Functions

- Perform arithmetic operations on polynomials
- Understand the relationship between zeros and factors of polynomials
- Use polynomial identities to solve problems
- Rewrite rational expressions

Creating Equations⋆

- Create equations that describe numbers or relationships

Reasoning with Equations and Inequalities

- Understand solving equations as a process of reasoning and explain the reasoning
- Solve equations and inequalities in one variable
- Solve systems of equations
- Represent and solve equations and inequalities graphically

Functions

Interpreting Functions

- Understand the concept of a function and use function notation
- Interpret functions that arise in applications in terms of the context
- Analyze functions using different representations

Building Functions

- Build a function that models a relationship between two quantities
- Build new functions from existing functions

Linear, Quadratic, and Exponential Models⋆

- Construct and compare linear and exponential models and solve problems
- Interpret expressions for functions in terms of the situation they model

Trigonometric Functions

- Extend the domain of trigonometric functions using the unit circle
- Model periodic phenomena with trigonometric functions
- Prove and apply trigonometric identities

Modeling

Modeling is best interpreted not as a collection of isolated topics but in relation to other standards. Making mathematical models is a Standard for Mathematical Practice, and specific modeling standards appear throughout the high school standards indicated by a star symbol (⋆). The star symbol sometimes appears on the heading for a group of standards; in that case, it should be understood to apply to all standards in that group.

Geometry

Congruence

- Experiment with transformations in the plane
- Understand congruence in terms of rigid motions
- Prove geometric theorems
- Make geometric constructions

Similarity, Right Triangles, and Trigonometry

- Understand similarity in terms of similarity transformations
- Prove theorems involving similarity
- Define trigonometric ratios and solve problems involving right triangles
- Apply trigonometry to general triangles +

Circles

- Understand and apply theorems about circles
- Find arc lengths and areas of sectors of circles

Expressing Geometric Properties with Equations

- Translate between the geometric description and the equation for a conic section
- Use coordinates to prove simple geometric theorems algebraically

Geometric Measurement and Dimension

- Explain volume formulas and use them to solve problems
- Visualize relationships between two-dimensional and three-dimensional objects

Modeling with Geometry⋆

- Apply geometric concepts in modeling situations

Statistics and Probability⋆

Interpreting Categorical and Quantitative Data

- Summarize, represent, and interpret data on a single count or measurement variable
- Summarize, represent, and interpret data on two categorical and quantitative variables
- Interpret linear models

Making Inferences and Justifying Conclusions

- Understand and evaluate random processes underlying statistical experiments
- Make inferences and justify conclusions from sample surveys, experiments and observational studies

Conditional Probability and the Rules of Probability

- Understand independence and conditional probability and use them to interpret data
- Use the rules of probability to compute probabilities of compound events in a uniform probability model

Using Probability to Make Decisions +

- Calculate expected values and use them to solve problems
- Use probability to evaluate outcomes of decisions

Conceptual categories, headings, and description of modeling quoted from Common Core State Standards.

The Common Core State Standards
for Mathematical Practice

The Standards for Mathematical Practice describe varieties of expertise that mathematics educators at all levels should seek to develop in their students. These practices rest on important "processes and proficiencies" with longstanding importance in mathematics education. The first of these are the NCTM process standards of problem solving, reasoning and proof, communication, representation, and connections. The second are the strands of mathematical proficiency specified in the National Research Council's report *Adding It Up*: adaptive reasoning, strategic competence, conceptual understanding (comprehension of mathematical concepts, operations and relations), procedural fluency (skill in carrying out procedures flexibly, accurately, efficiently and appropriately), and productive disposition (habitual inclination to see mathematics as sensible, useful, and worthwhile, coupled with a belief in diligence and one's own efficacy).

1. Make sense of problems and persevere in solving them.

Mathematically proficient students start by explaining to themselves the meaning of a problem and looking for entry points to its solution. They analyze givens, constraints, relationships, and goals. They make conjectures about the form and meaning of the solution and plan a solution pathway rather than simply jumping into a solution attempt. They consider analogous problems, and try special cases and simpler forms of the original problem in order to gain insight into its solution. They monitor and evaluate their progress and change course if necessary. Older students might, depending on the context of the problem, transform algebraic expressions or change the viewing window on their graphing calculator to get the information they need. Mathematically proficient students can explain correspondences between equations, verbal descriptions, tables, and graphs or draw diagrams of important features and relationships, graph data, and search for regularity or trends. Younger students might rely on using concrete objects or pictures to help conceptualize and solve a problem. Mathematically proficient students check their answers to problems using a different method, and they continually ask themselves, "Does this make sense?" They can understand the approaches of others to solving complex problems and identify correspondences between different approaches.

2. Reason abstractly and quantitatively.

Mathematically proficient students make sense of quantities and their relationships in problem situations. They bring two complementary abilities to bear on problems involving quantitative relationships: the ability to *decontextualize*—to abstract a given situation and represent it symbolically and manipulate the representing symbols as if they have a life of their own, without necessarily attending to their referents—and the ability to *contextualize*, to pause as needed during the manipulation process in order to probe into the referents for the symbols involved. Quantitative reasoning entails habits of creating a coherent representation of the problem at hand; considering the units involved; attending to the meaning of quantities, not just how to compute them; and knowing and flexibly using different properties of operations and objects.

3. Construct viable arguments and critique the reasoning of others.

Mathematically proficient students understand and use stated assumptions, definitions, and previously established results in constructing arguments. They make conjectures and build a logical progression of statements to explore the truth of their conjectures. They are able to analyze situations by breaking them into cases, and can recognize and use counterexamples. They justify their conclusions, communicate them to others, and respond to the arguments of others. They reason inductively about data, making plausible arguments that take into account the context from which the data arose. Mathematically proficient students are also able to compare the effectiveness of two plausible arguments, distinguish correct logic or reasoning from that which is flawed, and—if there is a flaw in an argument—explain what it is. Elementary students can construct arguments using concrete referents such as objects, drawings, diagrams, and actions. Such arguments can make sense and be correct, even though they are not generalized or made formal until later grades. Later, students learn to determine domains to which an argument applies. Students at all grades can listen or read the arguments of others, decide whether they make sense, and ask useful questions to clarify or improve the arguments.

4. Model with mathematics.

Mathematically proficient students can apply the mathematics they know to solve problems arising in everyday life, society, and the workplace. In early grades, this might be as simple as writing an addition equation to describe a situation. In middle grades, a student might apply proportional reasoning to plan a school event or analyze a problem in the community. By high school, a student might use geometry to solve a design problem or use a function to describe how one quantity of interest depends on another. Mathematically proficient students who can apply what they know are comfortable making assumptions and approximations to simplify a complicated situation, realizing that these may need revision later. They are able to identify important quantities in a practical situation and map their relationships using such tools as diagrams, two-way tables, graphs, flowcharts and formulas. They can analyze those relationships mathematically to draw conclusions. They routinely interpret their mathematical results in the context of the situation and reflect on whether the results make sense, possibly improving the model if it has not served its purpose.

5. Use appropriate tools strategically.

Mathematically proficient students consider the available tools when solving a mathematical problem. These tools might include pencil and paper, concrete models, a ruler, a protractor, a calculator, a spreadsheet, a computer algebra system, a statistical package, or dynamic geometry software. Proficient students are sufficiently familiar with tools appropriate for their grade or course to make sound decisions about when each of these tools might be helpful, recognizing both the insight to be gained and their limitations. For example, mathematically proficient high school students analyze graphs of functions and solutions generated using a graphing calculator. They detect possible errors by strategically using estimation and other mathematical knowledge. When making mathematical models, they know that technology can enable them to visualize the results of varying assumptions, explore consequences, and compare predictions with data. Mathematically proficient students at various grade levels are able to identify relevant external mathematical resources, such as digital content located on a website, and use them to pose or solve problems. They are able to use technological tools to explore and deepen their understanding of concepts.

6. Attend to precision.

Mathematically proficient students try to communicate precisely to others. They try to use clear definitions in discussion with others and in their own reasoning. They state the meaning of the symbols they choose, including using the equal sign consistently and appropriately. They are careful about specifying units of measure, and labeling axes to clarify the correspondence with quantities in a problem. They calculate accurately and efficiently, express numerical answers with a degree of precision appropriate for the problem context. In the elementary grades, students give carefully formulated explanations to each other. By the time they reach high school they have learned to examine claims and make explicit use of definitions.

7. Look for and make use of structure.

Mathematically proficient students look closely to discern a pattern or structure. Young students, for example, might notice that three and seven more is the same amount as seven and three more, or they may sort a collection of shapes according to how many sides the shapes have. Later, students will see 7×8 equals the well remembered $7 \times 5 + 7 \times 3$, in preparation for learning about the distributive property. In the expression $x^2 + 9x + 14$, older students can see the 14 as 2×7 and the 9 as $2 + 7$. They recognize the significance of an existing line in a geometric figure and can use the strategy of drawing an auxiliary line for solving problems. They also can step back for an overview and shift perspective. They can see complicated things, such as some algebraic expressions, as single objects or as being composed of several objects. For example, they can see $5 - 3(x - y)^2$ as 5 minus a positive number times a square and use that to realize that its value cannot be more than 5 for any real numbers x and y.

8. Look for and express regularity in repeated reasoning.

Mathematically proficient students notice if calculations are repeated, and look both for general methods and for shortcuts. Upper elementary students might notice when dividing 25 by 11 that they are repeating the same calculations over and over again, and conclude they have a repeating decimal. By paying attention to the calculation of slope as they repeatedly check whether points are on the line through $(1, 2)$ with slope 3, middle school students might abstract the equation $(y-2)/(x-1) = 3$. Noticing the regularity in the way terms cancel when expanding $(x - 1)(x + 1)$, $(x - 1)(x^2 + x + 1)$, and $(x - 1)(x^3 + x^2 + x + 1)$ might lead them to the general formula for the sum of a geometric series. As they work to solve a problem, mathematically proficient students maintain oversight of the process, while attending to the details. They continually evaluate the reasonableness of their intermediate results.

Connecting the Standards for Mathematical Practice to the Standards for Mathematical Content

The Standards for Mathematical Practice describe ways in which developing student practitioners of the discipline of mathematics increasingly ought to engage with the subject matter as they grow in mathematical maturity and expertise throughout the elementary, middle and high school years. Designers of curricula, assessments, and professional development should all attend to the need to connect the mathematical practices to mathematical content in mathematics instruction.

The Standards for Mathematical Content are a balanced combination of procedure and understanding. Expectations that begin with the word "understand" are often especially good opportunities to connect the practices to the content. Students who lack understanding of a topic may rely on procedures too heavily. Without a flexible base from which to work, they may be less likely to consider analogous problems, represent problems coherently, justify conclusions, apply the mathematics to practical situations, use technology mindfully to work with the mathematics, explain the mathematics accurately to other students, step back for an overview, or deviate from a known procedure to find a shortcut. In short, a lack of understanding effectively prevents a student from engaging in the mathematical practices.

In this respect, those content standards which set an expectation of understanding are potential "points of intersection" between the Standards for Mathematical Content and the Standards for Mathematical Practice. These points of intersection are intended to be weighted toward central and generative concepts in the school mathematics curriculum that most merit the time, resources, innovative energies, and focus necessary to qualitatively improve the curriculum, instruction, assessment, professional development, and student achievement in mathematics.

Published Titles in This Series